Analog
VLSI
Design
Automation

VLSI CIRCUITS SERIES

Series Editor

Wai-Kai Chen

PUBLISHED TITLES

PSPICE and MATLAB® for Electronics: An Integrated Approach,
John O. Attia

VLSI Design,
M. Michael Vai

Analog VLSI Design Automation,
Sina Balkır, Günhan Dündar, and Selçuk Öğrenci

Analog
VLSI
Design
Automation

Sina Balkır
Günhan Dündar
A. Selçuk Öğrenci

CRC PRESS

Boca Raton   London   New York   Washington, D.C.

Library of Congress Cataloging-in-Publication Data

Balkir, Sina
    Analog VSLI design automation / Sina Balkir, Günhan Dündar, A. Selçuk Ögrenci.
        p.   cm. (VSLI circuits series)
    Includes bibliographical references and index.
    ISBN 0-8493-1090-3
    1. Integrated circuits--Very large scale integration--Design and construction. I. Dündar, Günhan, 1959- II. Ögrenci, A. Selçuk (Arif Selçuk) III. Title. IV. Series.
TK7874.75.B35 2003
621.39'5—dc21
                                                               2003046211

Visit the CRC Press Web site at www.crcpress.com

© 2003 by CRC Press LLC

No claim to original U.S. Government works
International Standard Book Number 0-8493-1090-3
Library of Congress Card Number 2003046211
Printed in the United States of America 1 2 3 4 5 6 7 8 9 0
Printed on acid-free paper

To Sinem, Esra, Pınar, Seher, Oğuz, our parents,
and the North Aegean

Preface

The integrated circuit (IC) market growth has been explosive over the last decades. The increasing complexity of integrated circuits has been trying to follow market demands, which seem to expand exponentially year after year. The design of complex ICs has been possible due to two major factors: technology improvements and design automation tools. However, the above observations have been true largely for the digital VLSI domain. In the analog domain, designs have still been limited to a few hundred transistors at most. The most common and probably the only widespread design automation tool has been SPICE and its variants, which have in essence been the same for the last 30 years.

We believe that we are at the beginning of a new era in which analog design will be automated to a great extent. The design automation of analog integrated circuits is inevitable if one considers recently emerging disciplines such as System-On-a-Chip (SOC) design and Intellectual Property (IP) reuse. The fact that analog VLSI design automation tools have not penetrated into the market is due to several reasons in our opinion. The first one is that a general analog design automation methodology has not been defined; thus making it impossible to develop tools for various stages of the design process that will interact in a predefined manner with each other. The second reason is the difficulty of automating the process in analog domain. In our opinion, another major reason for the lack of mainstream analog design automation tools has been the resistance shown by the industry to such tools, which has been understandable due to the poor performance of earlier analog design automation tools.

In this book, we attempt to achieve several objectives. First, we try to define an analog VLSI design automation flow where every tool has its predefined objectives and interfaces. Next, we try to present working examples for each tool. A literature survey describing similar tools is included in the relevant sections of the book. Finally, we try to convince the reader on the validity of our approach by running our design automation system from top to bottom level for three different automation cases. These are definitely not the only examples in the book. Quite the contrary, examples pertaining to each tool are provided in the relevant chapters as well.

The book consists of six chapters. The first gives a general introduction, posing questions for the whole design automation flow, citing and comparing previous attempts for design automation tools and providing motivation for the subsequent chapters. Chapter 2 concerns itself basically with the system level design automation issues. System level design of three different systems is discussed in some length in this chapter. Furthermore, one of the most critical building blocks of the design automation flow, namely the performance estimator, is discussed in this chapter.

Chapter 3 concentrates on circuit level design automation, whereas Chapter 4 covers layout level design automation. Also introduced in Chapter 4 is another software, which we call the layout advisor. Top to bottom designs for three examples are presented in Chapter 5. We believe that these examples not only prove our point that design automation is possible in the analog domain, but also provide insight into how the design automation flow can be adjusted to different situations. Finally, Chapter 6 presents conclusions – although we believe that this field of discipline is far from being mature and that our work barely scratches the surface.

Acknowledgments

We are indebted to the following people for their contributions: Özsun Sönmez, for running all examples, modifying software to run in a streamlined design automation flow, the development of the switched capacitor design automation software, and improving the circuit synthesizer. Faik Başkaya and Balkır Kayaaltı, for their work in the performance estimator. Güner Alpaydın, for the circuit level design automation software. İ. Gökhan Erten, for the early versions of the switched capacitor design automation software. Selçuk Talay, for the ADC design automation software. Altuğ Şimşek, for ALG (Analog Layout Generator). M. Selçuk Ataç, for developing the layout advisor. Finally, Olcay Korol, for reading the manuscript.

The authors would like to thank the following institutions: *Department of Electrical Engineering of University of Nebraska-Lincoln* for providing prototype IC design and fabrication support, and test facilities that helped bring the circuit level design automation chapter of the book to fruition. We also thank the *Turkish Council of Scientific and Technological Research* (TÜBİTAK) for funding the ADC design automation section of the project and for the ADC test chips under project number EEAG-101E039.

Finally, Günhan Dündar would like to express his deep appreciation to the Electrical and Electronic Engineering Department of Boğaziçi University and General Electronics Laboratory (LEG) of The Swiss Federal Institute of Technology Lausanne (EPFL) for allowing him to devote the time necessary for this book.

About the Authors

Sina Balkır received his B.S. degree in Electrical Engineering from Boğaziçi University, İstanbul, Turkey in 1987. He received his M.S. and Ph.D. degrees in Electrical Engineering from Northwestern University, Evanston, Illinois in 1989 and 1992, respectively. Between August 1992 and August 1998, he was with the Department of Electrical and Electronics Engineering, Boğaziçi University as an assistant and associate professor. Currently, he is with the Department of Electrical Engineering, University of Nebraska-Lincoln, USA. His research interests include CAD of VLSI systems, Analog VLSI Design Automation, and Focal Plane Computation Arrays for Image Processing.

Günhan Dündar was born in İstanbul, Turkey in 1969. He obtained his B.S. and M.S. degrees from Boğaziçi University, İstanbul, Turkey in 1989 and 1991, respectively and Ph.D. degree from Rensselaer Polytechnic Institute, Troy, New York in 1993, all in Electrical Engineering. Since 1994, he has been with Boğaziçi University, where he is currently a professor. Between August 1994 and December 1995, he was with the Turkish Navy at the Naval Academy. Between September 2002 and May 2003, he was with EPFL, Switzerland, on sabbatical leave from Boğaziçi University. His research interests include CAD for VLSI, Neural Networks, and VLSI Design.

Arif Selçuk Öğrenci received his B.S. degree in Electrical and Electronics Engineering and in Mathematics, and his M.S. and Ph.D. degrees in Electrical and Electronics Engineering from Boğaziçi University, İstanbul, Turkey in 1992, 1995 and 1999, respectively. From 1992 to 1999, he was a research assistant. Currently, he is an assistant professor in the Electronics Engineering Department at Kadir Has University, İstanbul, Turkey. His research interests include Analog Design Automation, Neural Computation, and Analog Neural Networks.

List of Figures

List of Tables

Contents

Chapter 1

Analog VLSI Design Automation

1.1 Introduction

With the rapid advance in IC manufacturing technology, the IC industry has shown a boom unprecedented in history. From a single transistor in the 1950s to ICs containing millions of transistors in the beginning of 21st century, the improvement in IC complexity and functionality has been exponential. This growing complexity has brought along design problems as well; an ever increasing number of designers are employed by IC providers. However, it has not been enough to simply hire more engineers, but it has been necessary to increase designer productivity. This has been possible only by the prevalent use of Computer-Aided Design (CAD) tools.

Many CAD tools for design automation exist in the literature. These tools are either university-based research tools or tools that can be acquired through well-established companies. Moreover, their common point is that they are all intended for use within a well-defined design flow. This design flow has shown few variations between companies over the last decade. Thus, phrases like Automatic Test Pattern Generation (ATPG), behavioral simulation, channel routing, synthesis, etc. have become well-defined terminology in digital IC design. When new tools are developed or existing tools are improved, they are expected to fit into this design flow, which is often summarized in terms of a Y-chart, as shown in Figure 1.1.

When we think about the analog design domain, we notice that such a design flow does not exist despite the recent efforts of some researchers. Thus, the research on analog design automation, which has been very limited as opposed to digital design automation, has concentrated mostly either on specific tools designed for specific applications, or on tools targeted toward specific aspects of a design. The best-known example of the latter case is SPICE and its many variants. SPICE is an excellent tool for circuit-level simulation, but where it fits in a design flow is an open question. Many examples from the literature will be cited for the former case in later chapters of this book.

The major reason underlying this lack in analog design automation tools has been the difficulty of the problem, in our opinion. Design in the analog domain requires creativity because of the large number of free parameters and the sometimes obscure interactions between them. Traditionally, analog

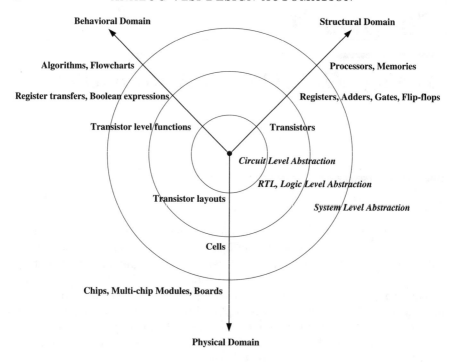

FIGURE 1.1
Y-chart showing design domains and abstraction levels.

designs have been carried out by experienced designers whose only aid is SPICE for simulation. These designers have not been able to quantify their knowledge in a formal framework suitable for CAD tools . Thus, analog design has remained more of an "art" than a "science".

The situation described above cannot continue in this direction in the future for several reasons. A short time-to-market is a big pressure on all designers. Using modern computers and Design Automation (DA) software, digital designers are able to design multi-million transistor chips in less than a year, whereas a team of analog designers may require a similar time to design an analog chip (or an analog section of a chip) consisting of a few hundred transistors. Furthermore, when the technology changes, the digital designers can update or adapt their designs in a matter of weeks, whereas the analog designers may require a time similar to the original design time.

In recent years, the IC Design industry has been moving to System-On-Chip (SOC) designs, where a single chip may contain both analog and digital circuits. For example, an SOC aimed at processing information acquired through radio transmission and acting on it may require an RF receiver, A/D converter, a microprocessor, a RAM block, a ROM block, digital circuitry for connecting to the ethernet, a D/A converter, filters and even some type of actuators. Typically, a designer, or even a company, cannot be an expert on all these different fields and may choose to use the designs of others. This brings us to the concept of intellectual property (IP) reuse in IC Design. IP reuse and the associated procedures have been around for the last 10 years and are slowly becoming an established discipline for digital design where the IP consists of a behavioral definition that can be synthesized targeting the end-user's technology. However, the same idea cannot be applied for analog design, where changes in technology may even dictate different circuits for the same application. To apply IP reuse in the analog domain, an Analog Design Automation (ADA) environment must be defined.

In this book, we propose a general methodology that can be used for many different applications. The design flow is created such that each tool is well-defined in terms of its duties, its inputs, and its outputs. The tools presented in subsequent chapters are state-of-the-art tools developed over the years by our research group, but new tools conforming to the definition of the design flow can be developed and these tools can replace our own. Furthermore, the design flow has been defined such that each tool is also useful by itself and the designer may choose to use them independently. The designer may also intervene at any point, choosing to bypass some of the tools and use his/her own intuition. As the reader proceeds through the book, he/she should keep this philosophy in mind.

We also illustrate our approach on three examples in this book. These examples are taken from the verbal definition phase to the physical layout generation phase. Their results are compared with the initial verbal specifications. The design examples chosen are Switched Capacitor (SC) filters, A/D converters, and Analog Neural Networks. SC filters are chosen to illustrate the performance of our system on sampled-data systems. A/D converters are the most common example of mixed-signal systems. Analog Neural Networks provide an example of continuous-time analog arithmetic systems.

In the next section of this chapter, a brief literature survey on complete analog CAD frameworks and terminology will be presented. Section 1.3 will describe our design flow and the associated design tools in our design flow. Section 1.4 concluding this chapter will outline the three examples that will be pursued throughout the book. Also, a very brief outline of the remaining chapters of this book can be found in this section.

1.2 Previous Analog Design Flows

A complete and rather recent overview on analog CAD can be found in Reference [1]. The authors review over a hundred different works on analog CAD ranging from macromodeling at the behavioral level to layout at the lowest level. They also outline a hierarchical design methodology similar to ours in many respects. Their methodology consists of a top-down synthesis and a bottom-up verification at every level of the hierarchy. Starting from a system concept, they first carry out their system design followed by simulation and verification. The system design is then given as input to the architectural design phase, which is the input to the cell design, cell layout, and system layout phases. Each of these phases has a feedback loop consisting of simulation and verification as shown in Figure 1.2. The top-down direction is called forward progress, whereas the bottom-up phase is called backtracking and redesign.

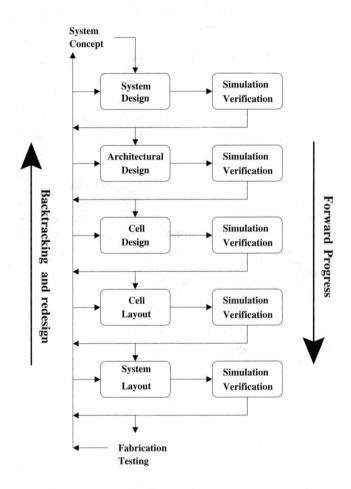

FIGURE 1.2
Analog IC design process.

Each block in the top-down direction takes an architecture or topology from the preceding block as well as performance constraints. Its output is a more detailed circuit diagram (or finally layout) as well as new specifications for the next block. Thus, each block in the forward direction does a specification translation and topology selection (Figure 1.3).

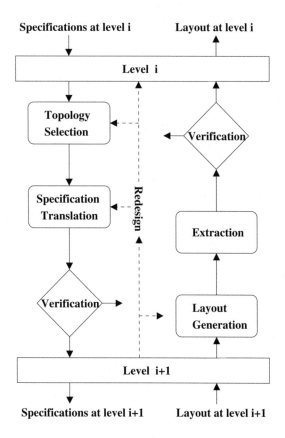

FIGURE 1.3
Hierarchical design strategy.

The advantage of propagating the constraints down the design hierarchy is that problems or inconsistencies in the specifications can be caught early on. Therefore, the ADA design flow requires tools for simulation at several levels of the design hierarchy, modeling at high levels, circuit synthesis, and layout generation. We will briefly review each of these in the next few paragraphs, leaving detailed description of existing tools to subsequent chapters.

One of the major tasks in ADA is simulation. Traditionally, this task has come to denote simulation at the circuit level and this is done numerically through SPICE. However, circuit simulation is not adequate or suited for many applications in ADA. For example, using traditional SPICE in a mixed signal circuit takes a huge amount of time where the digital sections could easily be simulated by a logic simulator. Also, simulating discrete time systems like SC filters or Σ-Δ modulators is time consuming, to say the least. For such systems, special numerical simulators must be developed. Another simulation application where SPICE is not suitable is in cases where the user would like to estimate the performance of a large circuit consisting of well-defined blocks to a certain accuracy. In such a case, using analytical models for the blocks forming the large circuit not only speeds up the simulation, but also gives the designer insight on how the overall circuit performance depends on a particular property of a certain block. Such models can be written in analog HDL's like AHDL or VHDL-AMS developed specifically for such applications. Another possibility for modeling blocks within a complex system is the use of macromodels, which are highly simplified circuits approximating the behavior of the block. For example, an opamp can be represented by a voltage controlled voltage source and a resistor. The advantage of using a macromodel is that a special simulator is not required, but SPICE will still do the simulation. One final application in ADA where simulation is required is in estimating the performance of a circuit before designing it. This is a very important problem for ADA that has not been satisfactorily solved and will be discussed in detail in the second chapter of this book.

Another type of simulation at the circuit level is symbolic analysis. Symbolic analysis is fundamentally different from numerical simulation, which tries to solve the circuit via nodal analysis by substituting element models into the nodal voltage matrix. The intermediate steps in a numerical simulator are meaningless to the human. On the contrary, symbolic analysis tries to derive equations for the circuit. For example, symbolic analysis will derive equations pertaining to the gain of an opamp based on the transconductance (g_m) and output impedance (g_o) of the transistors composing it. At first glance, this type of analysis seems very advantageous, as it makes the simulation very fast once the equations have been derived. Even more important, it gives the designer insight on how to design the circuit and which parameters are critical. However, symbolic analysis has several shortcomings. One of these is that DC analysis has to be performed for the transistor small signal parameters to be obtained. Thus, SPICE is still necessary for the solution. The second shortcoming is that the computer may generate huge expressions for medium complexity circuits that are not easy to simulate and much harder to interpret. Recently, several symbolic analyzers have been introduced, which can do simplification on these equations, even while generating them. However, symbolic analysis is not an alternative to SPICE analysis, but it complements it in some respects. It is an alternative to deriving equations by hand.

Another tool that is necessary for ADA is analog circuit synthesis and optimization. Analog circuit synthesis is the opposite of circuit simulation.

That is, an analog circuit synthesis tool tries to generate a circuit that meets some given specifications. In doing this, the tool performs mainly two tasks; topology selection and circuit sizing. Given a certain set of specifications, one must try to find the optimum topology within certain constraints while minimizing some cost. This is called topology selection and is the more difficult of these two problems. Some kind of library is necessary for topology selection. However, even with a library, it is not obvious as to which topology is going to yield the best results. For example, to attain a certain voltage gain, a two stage opamp may be enough. However, the transistor sizes of the two stage opamp may have to be so large to reach that gain that a three stage opamp with smaller transistors might yield a smaller area. Thus, selecting the suitable topology without transistor size information is very difficult. Most topology selection algorithms to date have been rather heuristic in nature and try to use the expert knowledge of the designer.

The second problem in analog circuit synthesis is circuit sizing. Here, the sizing and biases of all devices in a selected topology have to be determined according to the given specifications. This can be done either in a knowledge-based manner or by pure optimization. The older synthesizers were mostly knowledge-based, where the knowledge of how to optimize the topology was somehow encoded into the software. However, with the improvement in the performance of computers, optimization-based synthesizers started coming into play in the late 1980s. Optimization can be done either on equations or on simulations. In either case, the optimization problem is not a simple one and contains many local minima, thus requiring the utilization of special optimizers. Analog circuit sizing is one of the best-studied subjects in ADA and there is a plethora of literature on analog circuit sizing, which will be reviewed and classified in Chapter 3 of this book.

Layout synthesis is yet another tool that is required in an ADA environment. The earliest tools employed for analog layout synthesis were just the two dimensional counterparts of simple channel routers. Partitioning, placement and routing were applied to the circuit to obtain the final layout. However, it soon became obvious that analog layout had many properties not present in digital layouts. For example, transistors can be interdigitized with each other to obtain better matching. To obtain such layouts, specialized module generators were added into analog layout generation algorithms. However, this was not sufficient to obtain high performance designs. Performance constraints starting from the system level had been successfully translated to lower levels all the way down to the layout level, at which point they were not used any more, making the whole design automation process pointless. Thus, performance criteria started to be added to the layout level marking a shift in the analog layout generation paradigm, but also rendering the problem more difficult. Modern analog layout generators are performance- or cost-driven. A survey of analog layout generation literature and relevant details will be provided in Chapter 4 of this book.

Other useful tools that could be included in an ADA system are yield analysis tools, tools that calculate statistically the effects of device mismatches (like Monte Carlo analysis), and analog test pattern generation tools. All of these tools are still at their infancy stage and will not be covered in this book. However, we try to incorporate device matching in all stages of our design.

As mentioned above, ADA tools in the literature have been mostly geared toward a specific application (like SC filters, for example) or concentrate only on one level of the design. Until very recently, a tool suite that covers the whole ADA domain has not been available. Such a tool is now available [2]. This ADA system is called a "synthesis environment" by its authors as it incorporates many different tools around a single controller. The user who is designing a large circuit consisting of many blocks first has to select the block he/she wants to synthesize. In the AMGIE environment, this is called a "function block." A function block is typically a circuit like an OTA or an opamp. Next, the topology selection tool selects the appropriate topology. The topology is selected in a three step manner. In the first step, the absolute best performances of each block with respect to each parameter disregarding other parameters is listed. All the topologies remaining within the given constraints are selected whereas others are eliminated. Next, interval analysis is used to determine whether the remaining topologies still satisfy all constraints at the same time. The third filter ranks the remaining topologies for selection.

After completing topology selection, the user can either use the first ranking topology or can select another topology. Then he/she will have to use the sizing and optimization tool. The optimization tool is equation-based and uses DC equations derived from the circuit topology and AC equations obtained from symbolic analysis. Once the function block is designed, it enters the verification tool, which is actually a shell running a simulator according to given specs and tests whether simulation results conform to the specs. Once all function blocks have been generated, the system calls the layout generator, which is performance driven and generates layouts for the whole system. A final verification is run on the circuit extracted from layout using the verification tool mentioned above.

The AMGIE synthesis environment is capable of performing redesign iterations if specifications are not met. This is possible because optimization is done on equations and layout parasitics are only estimated before the layout is drawn. The redesign iterations are done according to the schedule built into the tool by expert designers. The tools are designed in a modular fashion so that the optimization algorithm or transistor models can easily be updated.

1.3 Proposed Design Flow and Tools

In this section, the design flow and tools proposed in this book will be introduced. A short discussion about each tool will be presented whereas the detailed descriptions can be found in subsequent chapters. The proposed design flow is depicted in Figure 1.4.

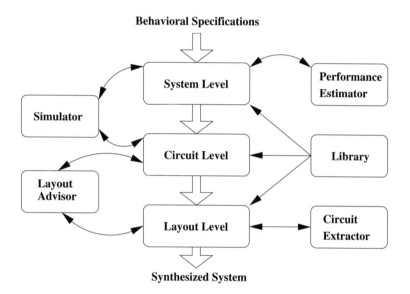

FIGURE 1.4
Proposed design flow.

1.3.1 System-Level Synthesis

The analog VLSI design system consists of eight major blocks as observed from Figure 1.4. The first block is the system-level synthesis block. It takes behavioral specifications as its input and provides a "block diagram" for the analog circuit interacting with the library and performance estimator closely. Although the main idea is the same, the system-level synthesis tool shows great variation from application to application. However, all of these tools possess common properties and have similar input and outputs. Hence, we still consider a system-level design block in our ADA flow.

As stated earlier, three examples will be discussed throughout this book, each with a distinct system-level tool. The SC system-level design tool uses filter specifications (like cut-off frequency, Q, etc.) as inputs. It runs a simple optimization based on ideal components to determine an initial guess for the order and type of filter. Then, it replaces the blocks (opamps and switches) with their macromodels, which include secondary effects for each block. The optimization is carried out again, this time incorporating power and area for the macromodels into the cost function. The power and area figures are taken from the performance estimator, which gives a quick estimate for each topology given various performance specifications. Topology selection for the blocks is also carried out as a byproduct of the system-level synthesis tool.

The second example is an analog neural network synthesis tool. Again, given the system-level requirements of the resulting neural network, the tool returns performance specifications for each block (which are the linearity and offsets for the synapse and neurons) and the weight values. Various synapse and neuron topologies can be selected from the library at this stage.

The final example is the ADC synthesis tool. Given performance specifications like linearity, resolution, speed, etc., the system-level synthesis tool first selects an appropriate architecture from among the three architectures modeled. These architectures are flash, pipelined ADC, and Σ-Δ converter. After the first step, which chooses the architecture, the block diagram is generated for that particular architecture depending on the components in the library and the performance required from the converter. The synthesis tools for flash and pipelined ADC's are available, however, at the time of writing the book, Σ-Δ converter synthesizer has been in progress. Hence, the subsequent chapters contain examples pertaining only to the first two types of ADC architectures.

It follows from the above discussion that all three applications require three different tools. If a new application is desired, a new tool has to be developed. This new tool should have the following properties:

- It should have a friendly user interface that a non-expert can use.

- It should run a fast optimization based on equations and/or macromodels for each block. These equations or macromodels can be implemented as a stand-alone software in a language like C or can be in a tool like VHDL-AMS.

- Its cost function should include costs for each block obtained from the performance estimator.

- It should have two outputs; a block diagram for the final system and performance parameters for each block.

As long as these rules are followed, new tools can be written for new applications, thus expanding the capabilities of the ADA system. It should also be noted that the ADA system was designed with the idea of each tool's being independent of each other and useful by itself. The system-level design tool can also be used as a stand-alone tool, which helps the engineer to explore issues and tradeoffs in analog system design. Many different runs of the system-level design tools will be presented throughout the book.

1.3.2 Circuit-Level Synthesis

The circuit-level synthesis tool used in the ADA system proposed in this book is optimization-based. It uses a simulator for calculating the DC solution and utilizes either user-defined equations or neuro-fuzzy performance models for the AC behavior. It also incorporates process variation and mismatch models that can be used for design centering.

The desired performance parameters as well as the selected topology for each block are supplied to the circuit-level synthesis tool. After optimization is carried out, the tool gives a sized netlist for the block in SPICE format. Again, with inputs and outputs defined as such, it can be used as a stand-alone tool for analog circuit designers.

The tool has several novel features, which will be discussed in detail and compared with other existing tools from the literature in Chapter 3. However, these features will be very briefly introduced in the next few paragraphs. Perhaps the most important aspect of the synthesis tool is its optimization algorithm. It is well known that the performance of simulated annealing (SA) is strongly dependent on its parameters like cooling regime, initial conditions, etc. A wrong choice of these parameters can result in the algorithm to be stuck at a local minimum. Unfortunately, these parameters change from example to example and there is no way of predicting them beforehand. Evolutionary strategies (ES) suffer much less from this problem. They tend to converge to a region around the desired solution, but may take a very long time or may even fail to converge to the final solution. Our algorithm combines these two and is shown to be superior in performance to both. A rigorous description of the algorithm and experimental results will be presented in Chapter 3.

Another strength of our circuit-level synthesis tool is that the DC points are determined by optimization over the DC simulator. Hence, there is no approximation in the DC solution. This may seem to be a disadvantage in speed, but it should be noted that any error in DC solution propagates to the later stages of the circuit optimization, as all parameters are dependent on the DC solution. The DC simulator used is a SPICE-like simulator developed specifically for such purposes, however, it has a significant speed advantage over ordinary SPICE simulation. The interested reader is referred to Chapter 3 for further discussion of the simulator.

AC simulation and optimization are carried out based on either hand-derived equations or neuro-fuzzy performance models. Using equations might seem like a waste of human effort, but the number of available topologies is limited, and an average designer can derive and verify these equations in less than a week. Once these equations have been derived, they can be used for that particular topology over a whole range of technologies. Furthermore, equations are very easy to be employed within the optimizer and result in fast optimization. Note that the equations can be supplied from a symbolic analyzer as well if one is available. The other option is to use neuro-fuzzy performance models. A script has been developed to that end. The script runs SPICE many times, parses the results and trains the AC performance without any human intervention. The disadvantage is that this process must be repeated for every technology change.

Another issue related to circuit synthesis tools in general is that they find a design that satisfies the given constraints, but the design is generally at a corner of the design space, thus resulting in a design that is very sensitive to process variations. To alleviate this problem and to achieve at least some sort of design centering, the effects of process variation were modeled and incorporated into the cost function of the optimizer.

1.3.3   Layout-Level Synthesis

The layout-level synthesis tool is a layout generator suitable for analog design. It is capable of reading hierarchical SPICE netlists and generating layouts in CIF format hierarchically. The analog layout generator (ALG) carries out three functions; namely, partitioning, placement, and routing. At the partitioning step, transistors that are connected to each other are grouped together. In the next step, an initial placement is obtained from the partitioner. Then, this placement is optimized iteratively. The final step is the router that runs on the placed netlist.

ALG is also capable of understanding simple directives given as commands in the netlist. These directives range from merging devices to placing them apart or placing some devices at specific locations of the layout. In its present form, this capability allows some control over the performance of the layout. While ALG cannot interpret any numerical directives pertaining to performance currently, a new performance-oriented version of ALG is under development. This version will be able to create layouts according to given performance parameters.

ALG can be used as a stand-alone tool that aids layout designers. The layout designer can also input some directives to ALG as explained above and has control over the final layout generated.

1.3.4   Performance Estimation

The performance estimator takes the topology and various performance parameters as input and gives various performance parameters as output. For example, it tries to answer questions such as, what is the power consumption and area of an opamp with a given gain and bandwidth. These answers should be given in a very short time since they will be used in the optimization loop of the system-level synthesis tool.

Performance estimation is one of the most crucial and difficult issues in ADA. There is no established methodology for this problem. Efforts in this direction have been various, ranging from dividing the block into subblocks and combining the estimates of each block to using interval arithmetic or even creating huge performance tables. A survey of existing techniques is provided in Chapter 2.

The performance estimator described in this book is based on building performance models using the associated models of basic blocks. Performance models for subblocks like transistor-resistor, transistor-current source, diff-amp, current mirror have been developed. The whole design space for these simple subblocks can be explored by the use of simple, yet accurate analytical models. Performance estimation for a building block such as an OTA is attained by combining the estimations for the subblocks in an intelligent manner, as will be explained in Chapter 2. The careful reader may note that the performance estimator is also useful as a stand-alone tool. It will aid the human designer in making decisions at the system level, whereby the designer can select a topology among many topologies or change the architecture of his system according to what type of blocks can be obtained.

1.3.5 Layout Advisor

The layout advisor takes a sized circuit netlist as input and creates "advice" for ALG. This is achieved by performing a sensitivity analysis on the circuit. Two types of constraints are generated by the layout advisor; namely, parasitic constraints and matching constraints. Layout advisor again has a SPICE engine for DC simulations and a specialized simulator for AC analysis. The layout advice could have been created within the circuit-level synthesis tool, but in accordance with the philosophy of stand-alone tools, this was not preferred. The output of the layout advisor is a circuit netlist, but it has extra comments in it that are advice for ALG. Sensitivity analysis will show which devices need careful matching and which parasitics limit the performance. Layout advisor will then present as output either qualitative advice that can be understood by the current version of ALG or quantitative advice giving limits on the matching and parasitics that can be interpreted by later versions of ALG. Similar to the tools described above, the layout advisor can be used as a stand-alone tool for giving advice to novice layout designers on important sections of the layout.

1.3.6 Library

The library interacts with all levels of design flow in order to maintain a consistent flow of information among different modules. Technology parameters of transistors, topologies of building blocks, design rules, and extraction rules are all compiled into this library.

1.3.7   Circuit Extractor and Simulator

These two final blocks have already become standard in analog design. The circuit extractor will extract SPICE netlists with parasitics from the layout. The resulting circuit will be simulated and verified against the desired performance parameters at the circuit level. The simulator is an external simulator. In later versions of ALG, the extractor will be integrated into the optimization loop of the layout generator. If the simulation results hold, then the circuit synthesis will have been successful. If the results are not within specifications, then the ADA system operator will have to change some specifications and resynthesize. In future versions of the ADA system, some intelligence will have to be incorporated into a central controller, which will make the retrace decisions.

It should be noted that the flow of the ADA system is controlled by a human operator. The operator can blindly run one software after another until the layout is obtained. However, the more experienced analog designer may not rely completely on the computer and may wish to intervene. It is the authors' belief that the ADA system design flow offers the operator the chance to become a designer by interpreting (and modifying if necessary) the intermediate data between the software. The designer can also run the software independently to get some idea about the individual sections of the design. In Chapter 5, several design examples will be carried out from the specification level to the chip level. The next section describes in more detail the three design examples covered throughout the book.

1.4   Design Examples

Three examples were chosen to illustrate the validity of the ADA approach presented in this book. The first example is a sampled data system; namely, a switched capacitor filter. The second example was chosen to illustrate an analog arithmetic circuit and is an analog neural network. The third example is the most common mixed signal circuit; that is, an ADC. In this section, we will review these three examples.

1.4.1   Switched Capacitor (SC) Filters

SC filters have been widely used in many areas of electronics since their introduction. However, the major issue in SC filter design has been the difficulty of simulating the designs. To be able to perform a SPICE analysis of the filter, a time domain analysis over many time periods has to be carried out for a single input frequency. Consequently, to obtain the frequency characteristics, many such simulations have to be done over a range of frequencies.

The first efforts toward design automation of SC filters has been in the direction of developing specialized simulators for SC filters. This is not an easy task, because such a simulator requires operation in the time, frequency, and z-domains. Over the years, such tools have become available [3]. After the establishment of simulators, these were used inside optimization tools to build synthesizers for SC filters such as the silicon compiler in [4]. Despite earlier efforts to automate SC filter design, the silicon compiler in [4] is the first fully automated design system of its sort. It contains a synthesis environment that performs LDI exact synthesis method for ladder filters. The synthesis environment interacts with both a numeric SC simulator and a symbolic simulator that operates in the z-domain. The silicon compiler also contains rather primitive tools for generating layouts. The predesigned cells are taken from a library and are routed together. The final verification of the design is done via electrical simulation through a fast circuit simulator which has become a commercial simulator over time.

The major drawback of all early SC synthesis tools has been the fact that the electrical simulation and the simulation used during synthesis have been different where secondary effects in switches and opamps have not been modeled. By using linear macromodels for opamps and switches, the influence of finite opamp gain, finite opamp output resistance, finite opamp output capacitance, and switch resistance have been computed in [5] for a variety of structures including the biquadratic cell (Figure 1.5). It may seem that these results are of limited use since new equations have to be derived for different structures. However, one should note that the biquad cell is a very versatile cell and can be converted into a low-pass, high-pass, band-pass, and band-stop filter by choosing capacitor values accordingly. Furthermore, biquad cells can be cascaded to obtain higher order filters. Although its component sensitivity and noise performance are worse than some other structures like the leapfrog structure, the modularity of the biquad led us to choose it as a building block for our structures. Our tool utilizes further improved versions of the models computed in [5] for optimization and hence minimizes the incompatibility between results of SC simulation and circuit simulation.

Optimization of SC filters using these macromodels were successfully illustrated in [6] and [7]. The details of system-level synthesis of SC filters will be provided in Chapter 2.

1.4.2 Analog Neural Networks

Neural networks have been successfully used in many different applications in the last two decades. For most of these applications, neural network software running on ordinary serial digital computers has been enough. However, for some applications, specialized neurocomputers have been required mostly due to speed limitations. Most of these neurocomputers have been parallel digital machines. Nevertheless, there has also been some research into the design of neural networks with analog circuitry. For a survey of neural network

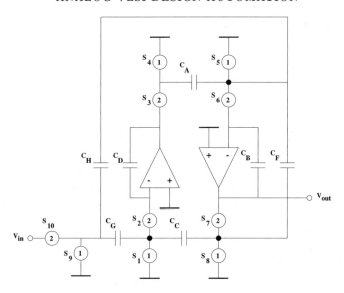

FIGURE 1.5
A biquadratic cell.

hardware, the interested reader is referred to [8]. Despite the fact that the analog implementation styles for neural networks may look to be the obvious choice at first sight, analog implementations have been plagued by a number of problems that have prevented them from going into mainstream use.

A simple multi-layer perceptron type neural network is shown in Figure 1.6. Here, the lines with arrows are the synapses that multiply the incoming signals by certain weights. The outputs of the synapses are added and passed through a nonlinearity at the neurons that are shown as circles in the figure. By changing the weights of the synapses, one can change the functionality of the network.

The first problem encountered in analog neural network hardware is the effects of limited precision in weights. For digital hardware, precision in weight storage is just a function of available area. With today's technology, precisions in the order of 16 bits are readily attainable. However, analog storage of weights is limited by many other factors depending on storage method and can rarely exceed 8 bits, which was shown to be sufficient for practical networks in [9].

The second major issue in analog circuitry is the fact that accurate operations cannot be carried out easily with small circuits. If one wants to implement a multiplication operation to an accuracy comparable to a 16-bit digital multiplier, one has to use a huge analog circuit, which defeats the purpose of using analog neural networks. On the other hand, using non-ideal synapses and neurons will cause errors that will accumulate at the output. The ef-

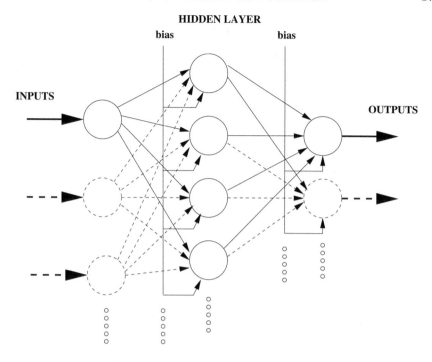

FIGURE 1.6
General structure of the perceptron.

fects of such errors on the overall performance of analog neural networks were modeled in [10]. The way to alleviate this problem is to incorporate these nonidealities into the training, which causes many extra local minima in the error surface and makes training difficult.

The third major problem in analog neural network design is process variations. These process variations will accumulate over synapses and neurons and will cause erroneous outputs. The minimization of this problem requires careful layout design and the availability of a well-controlled technology. Modeling and minimizing such effects was studied in [11].

In spite of all these problems, there have been successful efforts to design analog neural networks. Some of these have been through placing the training hardware on the chip as well, thus inherently including all problems into training [12]. If the training is successful, the resultant neural network will function as planned. Another research direction has been into modeling of all these secondary effects accurately in simulators used for training [11], [13]. In keeping with the philosophy of the ADA design flow introduced here, the system-level synthesizer will follow the second direction. More details on analog neural network system synthesis will be provided in Chapter 2.

1.4.3   Analog-to-Digital Converters (ADC)

One of the most difficult systems to synthesize is the ADC. The main reason is that it is a mixed signal circuit containing a large amount of circuitry. Another reason is that there are several different ADC types and each type has a plethora of architectures accumulated over many years of research.

In this book, we have decided to concentrate on two types of ADC's; namely, flash and pipelined, whereas Σ-Δ is still under development. The reason is that they provide a good cover of resolution and speed, as can be observed from Figure 1.7. For example, flash converters provide Gsamples/s conversion, but only for less than 8 bits, whereas more than 20 bits resolution is attainable with Σ-Δ converters at several hundred kHz.

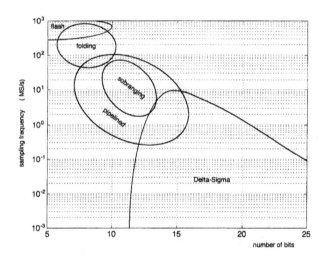

FIGURE 1.7
ADC domains.

There have been many efforts to model and synthesize ADC circuits. A good overview of CAD tools required for data conversion can be found in [14]. The authors not only present ADC synthesis tools developed up to that date, but also discuss the properties that must be present in an ADC synthesis tool. Out of the 11 architectures discussed in the paper, most of them can synthesize only one type of converter, while the most versatile can synthesize three different types. All of them are optimization-based and most generate their own layouts. Verification is done via dedicated behavioral simulation and/or formulas for most of them and only five out of the 11 tools discussed report working silicon. The authors stress that simulation for data converters is an issue and requires circuit-level simulation, mixed-signal simulation, macromodeling, and behavioral simulation simultaneously. They also mention that obtaining general equations for architecture selection is necessary, but sometimes very difficult to do. The careful reader may note here that all of the tools mentioned above are special-purpose ADC synthesis tools and are not intended for being plugged into an ADA flow.

Another tool developed earlier [15] does not try to synthesize an ADC from top level to layout level, but tries to discover the relationship between ADC specifications and the stage count and resolutions in a pipelined converter. The behavior derived here is one of the bases for our tool on pipelined converters. The results obtained are not valid for high-resolution pipelined converters where self calibration is utilized. For such architectures, a similar work that tries to optimize the stage resolutions is presented in [16]. These optimizations have just recently started to cover power consumption. Power optimization for a pipelined A/D converter is presented in [17].

These tools described in [15-17] have concentrated on pipelined converters. Similar tools have been developed or are under development for Σ-Δ converters. However, the design equations for such a converter are much more involved and the digital circuitry is also much more complex. The tool introduced in [18] approaches the problem at several levels. It uses equations at the architecture level, circuit simulation at the block level and has a specialized Σ-Δ behavioral simulator for verification. Silicon prototypes have been manufactured for two different examples. Again, the tool selects the architecture, calculates block specifications, selects the blocks, performs circuit optimization, and generates the layout. This tool comprises a specialized design flow from specification to layout and is not suitable for incorporation into a generalized ADA flow.

The remainder of the book is organized as follows: Chapter 2 discusses issues about system level synthesis and performance estimation in more detail as well as briefly reviewing macromodeling. Next, system-level synthesis tools for all three applications are extensively discussed along with several examples. Chapter 3 focuses on circuit-level synthesis and covers a detailed discussion of the optimization algorithm as well as mismatch and variation modeling in analog integrated circuits. This chapter also presents a brief summary of the extensive literature on this subject. Finally, a number of circuit synthesis examples are provided. Chapter 4 is mainly on the analog layout generation and layout advisor tools. Each step of the layout generation is illustrated with examples. Chapter 5 is intended to tie these concepts together and design examples are taken from the specification phase, carried out to the layout, and fully synthesized. The outputs at each stage are discussed along with the trade-offs. Finally, Chapter 6 concludes the book.

References

[1] Gielen, G. and Rutenbar, R., Computer-aided design of analog and mixed-signal integrated circuits, *IEEE Proceedings*, 1823, 88, 2000.

[2] Van der Plas, G. et. al., AMGIE-A synthesis environment for CMOS analog integrated circuits, *IEEE Transactions on CAD*, 1037, 20, 2001.

[3] Vandewalle, J., De Man, H., and Rabaey, J., Time, frequency, and z-domain modified nodal analysis of switched capacitor networks, *IEEE TCAS*, 186, 28, 1981.

[4] Assael, J., Senn, P., and Tawfik, M., A switched-capacitor filter silicon compiler, *IEEE JSSC*, 166, 23, 1988.

[5] Robertini, A. and Guggenbühl, W., Errors in SC circuits derived from linearly modeled amplifiers and switches, *IEEE TCAS – I*, 93, 39, 1992.

[6] Alpaydın, G., Erten, G., Balkır, S., and Dündar, G., Synthesis of switched capacitor filters in a multi-level optimization environment, *Proceedings of the Third International Workshop on Design of Mixed-Mode Integrated Circuits and Applications*, 175, 1999, Puerta Vallarta, Mexico.

[7] Alpaydın, G., Erten, G., Balkır, S., and Dündar, G., Multi-level optimization approach to switched capacitor filter synthesis, *IEE Proceedings – Circuits, Devices, and Systems*, 243, 147, 2000.

[8] Dündar, G. and Rose, K., Neural Chips, *Encyclopedia of Electrical Engineering*, John Wiley, 1998, 244, 14.

[9] Dündar, G. and Rose, K., The effects of quantization on multilayer neural networks, *IEEE Transactions on Neural Networks*, 1446, 6, 1995.

[10] Dündar, G., Hsu. F-C., Rose, K., Effects of nonlinear synapses on multilayer neural networks, *Neural Computation*, 939, 8, 1996.

[11] Öğrenci, A. S., Dündar, G., and Balkır, S., Fault tolerant training of neural networks in the presence of MOS transistor mismatches, *IEEE Transactions on Circuits and Systems*, 272, 48, 2001.

[12] Diotalevi F., Valle M., Bo, G. M., Biglieri, E., and Caviglia D. D., An analog on-chip learning circuit architecture of the weight perturbation algorithm, *Proc. ISCAS*, 419, 2000, Geneva, Switzerland.

[13] Bayraktaroğlu, İ., Öğrenci, A. S., Dündar, G., Balkır, S., and Alpaydın, E., ANNSyS: An Analogue Neural Network Synthesis System, *Neural Networks*, 325, 12, 1999.

[14] Gielen, G. and Franca, J., CAD tools for data converter design: An overview, *IEEE TCAS II*, 77, 43, 1996.

[15] Lewis, S. H., Optimizing the stage resolution in pipelined, multistage, analog-to-digital converters for video-rate applications, *IEEE TCAS II*, 516, 39, 1992.

[16] Goes, J., Vital, J. C., and Franca, J., Systematic design for optimization of high-speed self-calibrated pipelined A/D converters, *IEEE TCAS-II*, 1513, 45, 1998.

[17] Kwok, P. and Luong, H., Power optimization for pipeline analog-to-digital converters, *IEEE TCAS-II*, 549, 46, 1999.

[18] Medeiro, F., et al., A vertically integrated tool for automated design of Σ-Δ modulators, *IEEE JSSC*, 762, 30, 1995.

[19] Horta, N. and Franca, F., Algorithm driven synthesis of data conversion architectures, *IEEE TCAD*, 1116, 16, 1997.

Chapter 2

System-Level Design Automation

2.1 Introduction

As outlined in Chapter 1, an analog design automation system consists of three major feed-forward blocks and several smaller feedback blocks. The first major block is at the system level. As opposed to the other levels, system-level design automation is highly problem specific. This fact is revealed in the following sections, where several different systems will be studied. These systems are sampled-data analog filters, analog neural networks, and data converters. As discussed in the previous chapter, these systems are completely different from each other in nature. This difference reveals itself in the tools developed for system-level synthesis . However, the careful reader will observe that the design flow will converge to the same tools at the lower levels of the design process.

Although it was stated above that system-level design automation tools are highly problem-specific, they have many aspects in common. As a loose definition, the system-level design automation tool should take a system specification written for the system at hand by a person who is not an expert at CAD tools and create a block diagram for that system while, at the same time, converting the system specifications to a set of specifications for the circuit level tools below it in the hierarchy. This tool should take into account the properties of the system under consideration and also keep track of the trade-offs involved in the design process. The overall design must be optimum with respect to a set of criteria. As a result of this loose definition, it can be concluded that the system level design automation tool has the following major functions:

1. Take in a set of user specifications defined in a user-friendly manner.

2. Select a proper "block topology" for the desired system depending on the library at hand and the topology properties supplied to the tool by the tool designer. This topology might be the type and order of a filter for the first example in this book, the type and size of a neural network for the second example, or the type and architecture of the A/D converter for the third example.

3. Having chosen the topology, translate the system level specifications to constraints for each block.

Nevertheless, the selection of the block topology determines the constraints for each block. Consequently, the selected topology may yield unrealistic constraints for the blocks or be too demanding on the blocks as compared with another topology that might seem more undesirable in the beginning. A simple example could be a filter whereby the required specifications can be realized using a fifth-order or a seventh-order filter. The fifth-order filter might seem much more desirable in the beginning, but it might be so demanding on the active elements that the overall layout obtained with the fifth-order filter might be much larger in area as compared with the seventh-order filter. Therefore, the procedure outlined above is not a single run process, but can be iterative. At the same time, it is very slow to try to design the necessary circuits at each iteration and compare their performances. Instead, it is much more meaningful to develop a fast performance estimation tool that will run during the iterations and thus provide the system-level design tool with necessary information about each block. Performance estimation is discussed in Section 2.2 of this chapter.

Keeping the above considerations in mind, the following block diagram can be drawn for the system-level design automation tool.

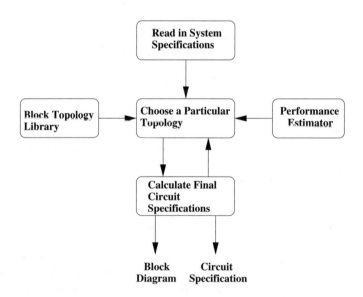

FIGURE 2.1
System-level design automation tool.

In Section 2.2, the concept of performance estimation and various implementations will be described. Section 2.3 will provide a short introduction to macromodeling of analog circuits. High level synthesis of switched capacitor filters is discussed in Section 2.4. The other two analog systems, namely neural networks and A/D conversion, will be the subject of Sections 2.5 and 2.6, respectively.

2.2 Performance Estimation

As described above, performance estimation is crucial to the operation of any analog synthesis system. Furthermore, it is also very useful as a stand-alone tool, as it can help the designer make choices about the architecture of the system. A more rigorous description of the performance estimation problem than the description in the previous section is as follows: Let y_i be the performance criteria of any analog circuit. For an opamp, these y_i can be gain, bandwidth, power, area, slew rate, etc. Let x_j be the independent parameters determining the performance. For a CMOS circuit, these are the transistor sizes (W and L) and bias voltages. Hence, we can define the performance of the circuit by the following i equations:

$$y_1 = f_1(x_1, x_2, \ldots, x_j),$$
$$y_2 = f_2(x_1, x_2, \ldots, x_j),$$
$$\vdots \qquad \qquad \vdots$$
$$y_i = f_i(x_1, x_2, \ldots, x_j). \qquad (2.1)$$

where f_i are the functions linking the performance parameters to the independent variables. These f_i are, in general, nonlinear functions and most of the time not explicitly written in analytic form. The problem is to find a way of expressing or at least estimating y_i in terms of each other without having to calculate x_j.

The difficulty of the problem is obvious from this definition. The simplest solution would be to form a lookup table for all combinations of x_j once and search from the table the required y_i combinations every time. This simple-minded approach suffers from several drawbacks. The first is the sheer size of the table. The boundaries on the W and L for the transistors are determined by the technology and the area requirements. Imagining that the lower bound is 0.8μm and the upper bound is 1000μm and the sizes are swept by 1μm steps, each W and L requires 1000 data points. For a rather simple opamp containing 10 transistors, the number of data points to be generated would be 1000^{20} ! Of course, the majority of the data points would be useless points with some transistors in the cut-off mode. However, the remaining data points would be too numerous to form a table where a fast search could be performed. Even if a way could be found around that problem, one will soon notice that there are duplicate solutions; that is, there would be several opamps with very similar gain, bandwidth, and power, but with different areas. Obviously, the solutions with large areas are superfluous and should be discarded, but the search for such solutions is prohibitively time consuming.

In [1], the authors have provided a solution to the above problems by creating the data points from a circuit-level synthesizer, namely OPTIMAN. Thus, they have avoided superfluous solutions. In 48 hours of CPU time on a powerful computer, they have generated 2500 samples, which, in turn, were filtered, yielding a total of 1853 useful samples. For their opamp, which contains over 20 transistors, this number is not enough to form a comprehensive table. The authors have instead opted for a neural network approximation to their data. Their examples demonstrate successful estimation of power and area from six performance parameters. The reader should note that a more comprehensive estimation problem involves the estimation of all performance parameters, thus necessitating a much higher number of samples.

To minimize the number of data points while, at the same time, to not sacrifice accuracy, the samples must be chosen very carefully. Harjani and Shao propose the combination of several methods to this end [2]. First, they try to define the feasibility region in order to limit the search space. This is done by generating the boundary points in the performance space by using vertical binary search. Then, they slice the volume dynamically to generate fewer data points for regression. They also screen the independent variables according to their significance on the performance parameters. The less significant parameters are sampled with much less frequency. Finally, they do a regression analysis on the data generated to form their models.

In [3], Veselinovic et al. propose to use interval analysis for performance estimation. Interval analysis had been applied to electronic circuits earlier [4], but the application had been limited to linear problems only. Interval analysis is a mathematical method whereby a linear set of functions of several variables (similar to Equation 2.1) can be used to estimate the intervals of the independent variable if the output variables are known. Our case consists of nonlinear equations, which can nevertheless be linearized regionally as shown in [3]. The problem that is being solved in [3] is not exactly the same problem as ours, but a related one in that the authors are trying to compare several topologies to select the most suitable one. In order to achieve this end, they try to estimate all performance parameters so that a comparison among all of them can be done.

Another analog performance estimation tool (APE) has also received much attention recently [5]. Although interesting in its own right, APE is not a performance estimator in our definition of the concept. On the contrary, APE tries to estimate the f_i in Equation 2.1. By doing this, APE can speed up the circuit optimization process. However, APE contains many interesting features that could be useful for the version of performance estimation introduced here. It is a hierarchical estimation engine where basic circuit elements such as MOS transistors, resistors, capacitors, etc., are modeled by symbolic equations. Basic analog building blocks such as differential amplifiers or current mirrors are also modeled symbolically, but at a higher level. The next level of modeling includes blocks like opamps. Thus, any large analog system can be modeled at several levels at the same time.

2.2.1 Estimation Methodology Used in the Book

Our performance estimator uses the same idea of basic analog building blocks and models the performance criteria analytically in terms of each other in closed form wherever possible. The estimation tool gets high-level design specifications such as bandwidth, gain, power consumption, etc. from the high-level synthesis tool as well as circuit topology.

The estimator tool first divides the complete circuit to blocks according to its device library. The device library contains analog building blocks for which analytical modeling has been done. Some of these blocks are:

- Common source with current source load

  - Current source as current mirror
  - Current source single biased transistor

- Differential pair with current mirror load

- Cascode

- BTS opamp

- Cascode opamp

The tool then distributes the performance specifications in an intelligent manner, and assigns them to the individual blocks. The next step is to find the area information using the performance specifications. Most analog blocks can be decomposed into driver and load sections such as a typical analog block depicted in Figure 2.2. For this figure, block 2 drives block 1. Examples for the driver can be "a common source driver transistor" or "a differential pair." A current source is an example for the load. Simple subblocks make modeling simpler and thereby reduce the computational efficiency.

Distributing the input specifications to subblocks 1 and 2 is not needed, since there is only one solution for a specific input combination. Subblocks are analytically modeled such that when given input parameters such as gain and bandwidth, the output will be transistor sizes and power consumption. The capacitances are calculated at the output nodes of each block; that is, only the output nodes of each block are considered for bandwidth calculation.

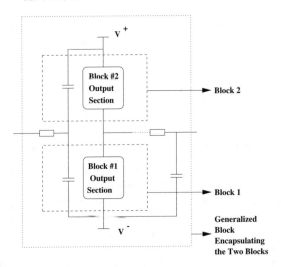

FIGURE 2.2
Typical analog subblock used for performance estimation.

The algorithm used is as follows: assume input specification for a block is entered as Gain and Bandwidth. Then,

$$GAIN = GM(driver\ block) \times R_{out}$$
$$= GM(driver\ block) \times (R_{driver\ block} || R_{load\ block}) \qquad (2.2)$$

The first step is to calculate R_{out} of block using the relation $R_{out} = 1/(\lambda\ i_d)$. Here λ represents the saturation drain current dependence on source to drain voltage (v_{ds}). Although this value depends also on v_{ds}; using the fact that this dependence is weak, a good starting approximation can be used. After GM has been found, transistor size information can be obtained.

$$GM(driver\ block) = f(Area\_driver\ block)$$
$$GAIN \times BANDWIDTH = GM(driver\ block)/(2\ \pi\ C_{out})$$
$$C_{out\ total} = C_{out}(block1) + C_{out}(block2) + C_{load}$$
$$C_{out}(block\ n) = f(constants + Area\_block)$$

C_{out} is the total output capacitance at the output node; (sum of output capacitances of the two blocks and a possible input capacitance of the next cascaded stage).

The contribution of each block to the output capacitance can be related to its area. To account for the reverse diode capacitances at the output node (which are very dependent on the reverse bias DC voltages); an algorithm is used to find the DC point of the output node according to the voltage-pulling strengths of the blocks.

Before the calculation of strengths, the output node is given an initial value condition since an initial transistor sizing is needed to commence calculations. The algorithm can calculate the strengths after the first run. A single iteration will be sufficient for getting a meaningful approximate voltage value for this node. This situation is shown in Figure 2.3. Here, $\beta = (K_p\,W/L)$ of a

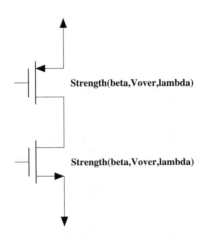

FIGURE 2.3
Drive strengths of transistors.

transistor. Using the channel length modulation, transistor strengths are employed to find the connection node voltage. The estimator also obtains the voltages of the nodes at the subblock connections, thus enabling power estimation.

2.2.2 A BTS Opamp Estimation Example

A typical BTS opamp consists of a differential amplifier gain stage and an output stage for high swing. In addition, frequency compensation is usually applied. Figure 2.4 is a simplified diagram of a BTS opamp.

Partitioning lies at the core of the performance estimation approach. Each large circuit is divided to smaller cells. For the BTS opamp example at hand, these are the input and output stages. Performances of each cell are evaluated separately. In order to simplify the problem more, some cells can be partitioned further, as shown with the dashed rectangles in Figure 2.4.

The performance estimator begins the evaluation from the output stage

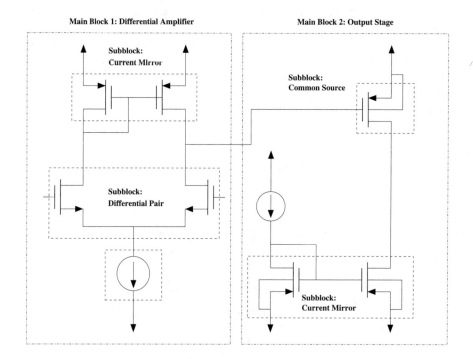

FIGURE 2.4
Uncompensated BTS opamp.

and moves on to previous stages after each stage is evaluated. To explain the operation, we will just consider two stages as mentioned above.

i. BTS output stage (common source with current mirror load):

The output stage of the BTS opamp is redrawn in Figure 2.5 for convenience. The output stage of an opamp is usually designed to have low output resistance in accordance with the definition of an ideal opamp. Another important factor is, of course, the gain that this stage will provide, as the circuit under consideration has just two stages. Thus, the output stage should contribute to the gain as much as possible. Other important considerations for the output stage are bandwidth and power consumption. Hence, we can state that the requirement space consists of r_{out}, *gain*, *bandwidth*, and *power consumption* of the block. The typical design parameters, on the other hand, are the gate drive voltage of the common source driver (V_g), the current (I_d), and the transistor sizes. Therefore, the design space consisting of these parameters should be explored in order to see which designs conform

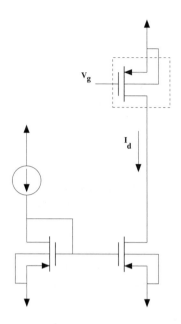

FIGURE 2.5
The common-source output section.

to the specifications. Calculating or assigning the V_g value is critical in that it forms a new requirement on the differential amplifier in that the differential amplifier design should set that value.

The output stage design exploration is achieved by the following formulae where G is the gain and r_{out} is the output resistance of the output stage.

$$G = g_m \ (r_p \parallel r_n) \qquad (2.3)$$
$$r_{out} = (r_p \parallel r_n)$$

Since G and r_{out} are provided as requirements, $(r_p \parallel r_n)$ and g_m can be calculated easily.

Another fixed requirement is the bandwidth given by,

$$BW = \frac{1}{(2\pi r_{out} C_{out})} \qquad (2.4)$$

where $C_{out} = C_{Load} + C_{transistors}$. Since bandwidth and the external capacitive load, C_{load} are also given, the estimator can also calculate the maximum total transistor capacitance at the output, $C_{transistors}$.

One useful information is the offset voltage of the output node, which is typically designed to be at the midpoint of two supply voltages. This value is 0V, when using symmetric supply voltages. The g_m value extracted from Equation 2.3 gives a size constraint of the driver transistor when working with a specific current. The current limit is provided by the power consumption figures. Equation 2.4 gives information about the sum of the transistor widths (W_p, W_n). The channel lengths of the transistors (L_n and L_p) are held at an interval whose lower limit is determined by the fabrication tolerances of the technology and upper limit by the performance requirements such as r_{out}.

Utilizing the above approach, candidate designs with varying transistor sizes (which translate to area), drain currents (which translate to power), and gate drive (which translates to a constraint for the previous stage) are calculated.

An example estimation run was performed on the output stage. The question was to estimate the area requirement for an output stage with a gain of 380, a power consumption of 5 mW from a +5V supply and an output resistance of 150kΩ. The estimator calculated a design with a gate area of around 4000 μm^2 minimum. Furthermore, it specified a gate voltage of 3.702V. The human designer given the same specifications obtained similar results working on SPICE. These results are shown in Table 2.1.

TABLE 2.1 Output Stage Estimation Results.

	SPICE	Estimator
Current (μA)	504	497
M1 (W/L)	1000μm/3μm	928.25μm/3μm
M1 (W/L)	400μm/3μm	380μm/3μm
V_g (V)	3.715	3.702

ii. BTS differential amplifier (differential pair with active load):

The first stage of the BTS opamp is the differential amplifier as shown in Figure 2.6. The differential amplifier block itself can be decomposed further into three subblocks. These are the current mirror composed of the transistors M3 and M4 in Figure 2.6, the differential amplifier composed of transistors M1 and M2, and the current source.

Since each arm carries a constant current when the inputs are equal, the output node voltage is adjusted by the size of the active current mirror. The output node voltage constraint was derived from the design of the next stage, as explained earlier. Again, the channel lengths are within an interval determined by the fabrication variance behavior of the technology and the performance criteria. Furthermore, channel lengths affect the output resistance directly. A similar approach was employed in the design of the differential pair

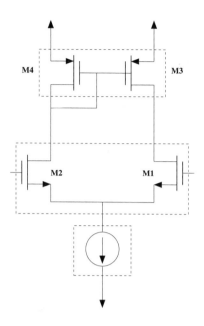

FIGURE 2.6
BTS Differential amplifier.

and similar agreement was obtained between estimator and human designer.

2.3 Macromodeling

For high-level synthesis applications, modeling of the building blocks is of utmost importance. Too simple models will cause loss of accuracy, whereas too complex models will not provide much speed advantage. The model must be complex enough to accurately represent all of the crucial secondary effects, but must be simple enough for the computer to operate on it very fast. Another important issue in many cases is the interpretability of the model.

The model used for high-level synthesis problems can be an analytical model as in the A/D converter example in this book or it might be a macromodel as in the switched capacitor example. Note that all the high-level models used in all the examples of this book, whether they are analytical models or macromodels, have been derived by humans. This has been a design decision on the part of the authors in that they wanted to keep the intelligibility of the models intact at all levels. However, this does not rule out addition of automatically generated macromodels as in [6] to the present examples or future examples. Automatic generation of macromodels involves not only the optimization of the component values within the macromodel to obtain the desired response, but also the optimization of the structure of the macromodel in many cases. Macromodeling is a rather involved subject that is not directly addressed on the design flow outlined in Chapter 1. The interested reader is referred to the plethora of existing literature.

2.4   High-Level Synthesis of Switched Capacitor Filters

It was realized very early on that switched capacitor filters are rather regular structures and thus amenable for design automation. As summarized in Chapter 1 of this book, the major drawback for such systems has been fast and accurate simulation. Such a simulation requires work in the frequency, time, and z domains. Earliest efforts for switched capacitor simulation systems go back more than 20 years [7]. Once a simulation methodology has been defined, optimizers working around these simulators can be built and used in the design automation of SC filters. One of the first examples in this field is the compiler described in [8]. This compiler is based on two simulators, a symbolic z-domain simulator and a numeric SC simulator. The final results are checked via circuit simulation. The main reason for so many simulations has been the difficulty of being able to model the nonidealities of the components inside the SC filters and the secondary effects associated with them. Thus, high-level simulators ignoring nonidealities are used for the initial design, whereas final fine tuning requires more exact simulations that cannot be used throughout the optimization loop due to speed constraints.

Modeling secondary effects in SC systems has not been ignored at all in the literature. One such effort [9] uses rather simple models for opamps and switches, but obtains very good agreement with practice. Here, opamps have infinite input impedance, finite gain and some output resistance. Switches are also modeled as having on resistances. These models were used in SC synthesizers recently [10], [11] with good results. Actually, the approach outlined in this book is an improved version of those synthesizers. In [10] and [11], ideal SC equations were used first to get an approximate solution in terms of C values and order. Next, more accurate models presented in [9] were substituted. Using the bilinear transform, complex frequency domain relations were obtained. Since these equations are not analytically solvable, an optimizer was used for the solution.

In this book, we present an improved version of this approach. It should be noted that the models used in [9–11] and the following pages are valid for the biquadratic filter (Figure 2.7). This figure is the same as Figure 1.5, but is replicated here for convenience. This is not a limitation because biquads can be cascaded to form higher order filters. The models of [9–11] are given below.

$$H(z) = \frac{n_0 z^2 + n_1 z + n_2}{q_0 z^2 + q_1 z + q_2}$$
$$n_0 = C_d C_h \alpha_2 (1 - \gamma_2)$$
$$n_1 = C_a C_g \alpha_1 \alpha_2 (1 - \gamma_2) + C_d C_h \alpha_2 [\gamma_2 - 1(1 - \gamma_2)\beta_1]$$
$$n_2 = C_a C_g \alpha_1 \alpha_2 \gamma_2 - C_d C_h \alpha_2 \beta_1 \gamma_2$$
$$q_0 = C_b C_d + C_d C_f \alpha_2 (1 - \gamma_2)$$
$$q_1 = C_a C_c \alpha_1 \alpha_2 (1 - \gamma_2) - C_b C_d (\beta_1 + \beta_2) + C_d C_f \alpha_2 [\gamma_2 - (1 - \gamma_2)\beta_1]$$
$$q_2 = C_a C_c \alpha_1 \alpha_2 \gamma_2 + C_b C_d \beta_1 \beta_2 - C_d C_f \alpha_2 \beta_1 \gamma_2$$

where α, β, and γ are the distortion quantities and the indices 1 and 2 stand for the left-side and right-side integrators, respectively.

The careful reader will also note that the nonidealities modeled above are only static nonidealities. Dynamic nonidealities are not taken directly into account. However, it will be seen that some dynamic effects are indirectly addressed in the discussion below.

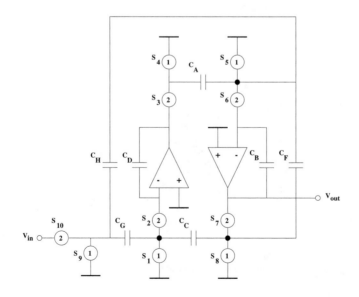

FIGURE 2.7
A biquadratic cell.

The input specifications for the high-level synthesizer are the clock frequency, center frequency, and the Q factor of the filter. The pertaining synthesis flow encapsulates two levels of optimization driven by the popular simulated annealing technique. In the first stage of the optimization, the aim is to find the minimum order of the ideal filter that satisfies the performance criteria. Since nonidealities degrade the performance, the actual filter order will be equal to or greater than this order. In this stage, the optimized variables are C_a, C_b, C_c, C_d, and C_f of the biquad. Since C_h does not affect the ideal response and $C_g = 0$ for bandpass filters, they are not taken into account. At each iteration, one of the capacitances is chosen randomly and perturbed such that the maximum change is 20%. The new value is also checked to be between the upper and lower capacitance limits and to satisfy the C_{max}/C_{min} ratio. All of these limits are given by the user and read from the technology file. The program flow is as in figure 2.8. As can be observed from Figure

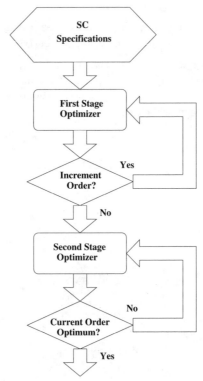

FIGURE 2.8
Flow chart of the algorithm.

2.8, this first optimization stage ends when the performance criteria are met; otherwise, the order is incremented. Due to the very simple equations, the evaluation of each solution is very fast. In the second stage, the best solution of the first stage is used as the initial solution. The optimized variables are the C_a, C_b, C_c, C_d, C_f, and C_h as well as the gain and r_{out} of the two opamps and r_{on} of the switches. The algorithm is similar with the following exceptions:

- At each iteration, r_{on} and r_{out} are checked in addition to the limits implemented in the first stage. $5r_{on}$(or r_{out})$C_{total} < T_{clk}/2$. This will prevent slew rate problems, assuming that the opamp has been designed such that it has no slew rate problems without any load.

- In the first stage, f_{center} and Q of the filter are calculated using approximate equations that are derived by the bilinear transform. In the second stage, the exact substitution for z ($z = e^{sT}$) is used. Novel equations obtained by this substitution give exact results and eliminate the need for iterations.

- In addition to f_{center} and Q differences from the desired values, the cost function also includes the area and power of the opamps and area of the switches that are calculated from A_0 (gain), r_{out} and r_{on} using the performance estimator.

- When the performance criteria are met, the optimization continues for higher order filters, searching whether a lower cost is possible with relaxed opamp and switch specifications. If there is no improvement after a few order increases, the optimization is finished.

The z-domain transfer function of the biquad with ideal elements was given above as:

$$H(z) = \frac{n_0 z^2 + n_1 z + n_2}{q_0 z^2 + q_1 z + q_2} \tag{2.5}$$

Substituting $z = e^{j\omega t}$ and applying some trigonometric identities, an expression for the magnitude response in terms of ω that is not too difficult to handle can be obtained. To find the center frequency, we take the derivative of this expression with respect to ω and equate it to zero. It is possible to obtain a closed form solution for ω in this manner. Hence, no iterations are necessary in contrast to the previous works cited above. To findthe Q, we first calculate the maximum magnitude; that is, the magnitude at the center frequency. Then, wedivide the squared magnitude by two and solve the resulting trigonometric equation again. Finally, the Q factor is calculated easily by

$$Q = \frac{\omega_c}{\omega_{cut1} - \omega_{cut2}} \tag{2.6}$$

A **Design Example:** One switched capacitor bandpass filter example is studied at this point to clarify the concepts presented above. The task is to synthesize a bandpass filter with center frequency at 10 kHz and Q of 40. After the first step of the synthesizer, which utilizes ideal opamps and switches, the results presented on the first three columns of Table 2.2 were obtained. The reader will observe that, although the optimization carried out

TABLE 2.2 Bandpass Filter Synthesis Results.

	Initial (Ideal Comp.)			Initial (Real Comp.)		Final (Real Comp.)	
	Bilinear	e^{sT}	Hspice	e^{sT}	Hspice	e^{sT}	Hspice
f_c	10006	9695	9700	9714	9600	9998	10030
Q	40	42	42	22	21	40	40
order		1			1		1

on the bilinear model reached the optimum point with 10006 Hz and a Q value of 40, SPICE simulation with ideal elements showed a center frequency of 9700 Hz and a Q of 42. The error in center frequency is unacceptable in many cases and results from the bilinear approximation that is being widely used in the literature, including [9–11]. However, if e^{sT} is substituted instead of z and the aforementioned simplifications are carried out, our f_{center} coincides with simulations to a very high accuracy. The next two columns demonstrate the effect of real components represented with macromodels (finite opamp gain and r_{out} and finite switch resistance r_{on}) on the performance of the SC filter. Again, the careful reader will notice a rather small variation on the center frequency, but a huge deterioration on the Q factor of the filter. The final two columns demonstrate the results of the high-level synthesis. Our synthesis stops with the center frequency at 9998 Hz and $Q = 40$. SPICE simulation results with macromodels demonstrate an error of only 0.3% on the center frequency. The evolution of the capacitance values and switch resistances is illustrated in Tables 2.3 and 2.4. These results are valid for $A_0 = 191$, $r_{out} = 21,423\Omega$, and $r_{on} = 17,428\Omega$.

TABLE 2.3 The Evolution of Capacitance Values (pF).

	C_a	C_b	C_c	C_d	C_f	C_g	C_h
initial	2.3	15.87	1.15	0.46	0.23	0	0.46 (chosen)
final	14.72	25.07	0.46	0.69	0.23	0	0.46

TABLE 2.4 The Evolution of Switch Resistances (Ω).

$S_{1,2}$	$S_{3,4}$	$S_{5,6}$	$S_{7,8}$	$S_{9,10}$
17,428	544	520	11,618	17,428

The evolution in frequency response of the same design is illustrated in Figure 2.9.

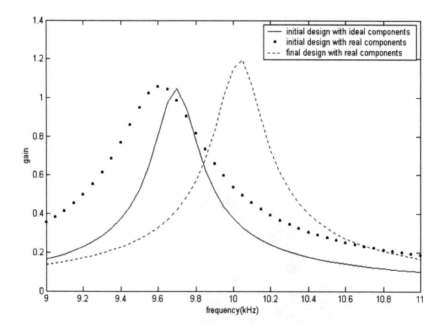

FIGURE 2.9
Frequency response of the design.

With our modeling and synthesis tool for SC filters, it is also of interest to observe the effects of various nonidealities on SC filter behavior. Figure 2.10 depicts the effect of A_0 and r_{out} on the center frequency. It can be seen that both A_0 and r_{out} have minor effect on the center frequency except for extreme values.

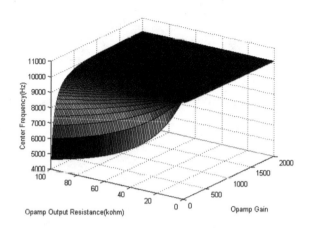

FIGURE 2.10
The effect of A_0 and r_{out} on the center frequency.

Figure 2.11 illustrates the variation of Q with the same parameters. Again, r_{out} is not very effective, but A_0 varies Q over a wide range.

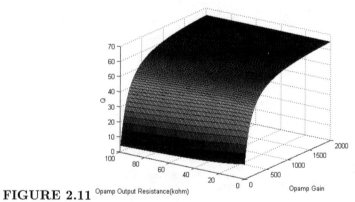

FIGURE 2.11
The variation of Q with the same parameters.

Figure 2.12 depicts the relation between f_{center} and r_{on} and Q and r_{on}. It is interesting to observe that both curves are highly nonlinear and have a minimum for r_{on} of approximately 25 kOhm.

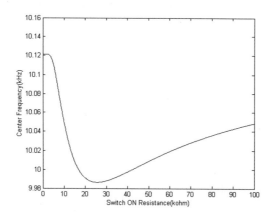

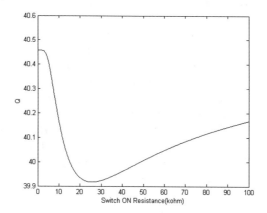

FIGURE 2.12

The relation between f_{center} and r_{on} and Q and r_{on}.

2.5   High-Level Synthesis of Analog Neural Networks

Efficient implementation of artificial neural networks in hardware is an important step toward the realization of artificial intelligence. Almost all intelligent algorithms have been developed for mimicking the behavior of the human brain by considering the following questions:

- How does the brain receive data from the outside world through sensors?

- How does the brain manipulate sensory data into information?

- How does the brain store and process the information to make assertions?

Studies on the structure of the human brain have revealed that the brain is composed of approximately 10^{11} processing nodes (neurons) connected massively via approximately 10^{15} synapses [12, 13]. This huge information processing network is known to operate based on pulses (electrical signals of magnitude in the order of 100 mV), where each neuron attains a relatively low activity level (10–100 spikes per second). However, the massive parallelism in the structure and the efficiency in power consumption result in a system that can perform six orders of magnitude more operations with eight orders of magnitude less energy with respect to a state-of-the-art computer system. Beyond the evident superiority of human neural networks in terms of processing power and energy, they are also capable of easily performing "difficult" tasks such as classification and recognition. Cognitive problems like spoken phoneme recognition or classification of handwritten digits require that the system (brain) should be able to learn from examples, which is not an easy task for a sequential software code implementing a certain algorithm since such problems usually do not possess well-defined algorithmic solutions. The structure of the brain as a neural network has inspired the design of artificial neural networks (ANN) that would mimic/simulate/emulate the interconnection structure and dynamics of human neurons so as to enable learning for such "difficult" tasks.

Based on the three main criteria given below, several types of ANN models have been formulated in the literature [14–16]:

- presence of a teacher during learning {supervised, reinforcement, unsupervised};

- type of learning algorithm {Hebbian, gradient descent-based, Winner-Take-All};

- type of connection structure {feedforward, feedback, bidirectional}.

In this book, feedforward neural networks with supervised, gradient descent-based learning are taken into consideration; however, the methodology can equally well be applied to other structures. The primary advantage in the realization of ANNs would be that they do not describe any algorithm for the direct solution of the problem at hand, whether it be a function approximation or classification problem. Rather, ANNs learn from examples, either supplied by a teacher indicating the correct solution (supervised), or the level of success (reinforcement), or supplied without any label (unsupervised). In the latter case, ANNs simply detect the inherent dynamics of the system from the sample data. The essential requirement for ANN is, however, that ANN should be able to function properly when new, previously unknown input data are applied to the network. This is the so called "generalization property" of ANNs.

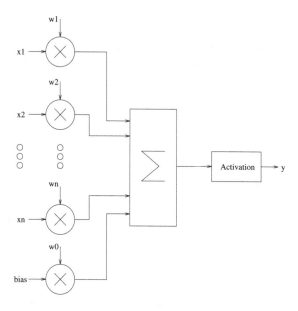

FIGURE 2.13
Structure of a neuron (single perceptron).

Formally, a neuron in ANN is a processing unit with several variable inputs $x_1, x_2, \cdots, x_n$ and (usually) one fixed bias (threshold) of value -1, denoted by x_0 (Figure 2.13). The output y of the neuron is a weighted sum of its inputs, passed through a (usually) nonlinear activation function, ϕ,

$$y(x_1, x_2, \cdots, x_n) = \phi\left(\sum_{i=0}^{n} w_i x_i\right) \tag{2.7}$$

where w_i is the weight associated with the input x_i. This type of a neuron is also called a perceptron. A feedforward, multilayer perceptron (MLP) network is a layered structure consisting of several neurons in each layer (Figure 2.14) where the external inputs are applied to the neurons in the first layer and, subsequently, outputs of each layer are fed to inputs of neurons in the next layer. Outputs of the last layer are the external outputs. Layers that do not have any direct connection to the external world are called hidden layers. Even though there is no theoretical limitation on the number of hidden layers in an MLP, usually a maximum of two hidden layers are used in practice. In this chapter, fully connected MLP structures are considered; that is, outputs of each neuron in any layer except the output layer are connected to each neuron in the next layer. The overall output of the MLP depends on the inputs and the weights. Given a training set $\mathcal{X} = \{\mathbf{x}^p, \mathbf{r}^p\}_p$, where $\mathbf{r}^p$ is the desired output vector corresponding to the input vector $\mathbf{x}^p$, learning for the MLP denotes determination of the optimal set of values for the weights so as to achieve an acceptable level of approximation to desired values. The convergence criterion is usually expressed in terms of the rms (root-mean-square) error for the training set given as,

$$E_{rms} = \sqrt{\frac{1}{pd} \sum_{i=1}^{p} \sum_{j=1}^{d} (\mathbf{r}_j^i - \mathbf{y}_j^i)^2} \qquad (2.8)$$

where d is the dimension of the output vector. Learning is based on applying the input samples sequentially to the MLP and calculating the outputs where the weights are initialized to small random numbers. Those outputs are compared with target values and the free parameters of the MLP (weights) are updated according to the gradient descent algorithm where sensitivity of output error with respect to each weight value is used in the update as follows,

$$\Delta w = -\eta \frac{\partial E}{\partial w} \qquad (2.9)$$

where $E = (r^p - y^p)$ is the error at output of a single neuron for the pth pattern and η is the learning rate. The essential principle underlying the backpropagation (BP) of the output error using gradient descent is that the error forms a hypersurface in the weight space and the global minimum of this surface (where all partial derivatives are zero) is the optimal solution if one exists. It should be kept in mind that the error surface may also possess local minima where the above condition is met, hence, the backpropagation algorithm usually stops at a local minimum depending on the initial weight set. Since the full mathematical treatment of the error surface for an MLP with hundreds of weights is very elaborate, it is of little concern if a local minimum satisfying the convergence criteria is reached instead of the global minimum. As there is no guarantee that the algorithm will converge to an acceptable level of error, it may be necessary to restart the process with a new set of initial weights and/or training parameters (e.g., learning rate, momentum term, decay factor) if the solution is not satisfactory.

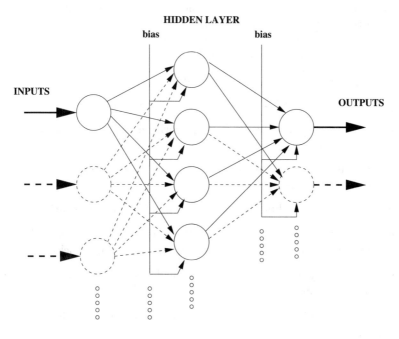

FIGURE 2.14
The general structure of the multilayer perceptron.

As models of information processing systems, ANNs have been implemented by software or hardware in a variety of disciplines for many purposes. High-energy physics (particle detection), engineering, financial markets (time series prediction, credit card fraud detection), medicine [17], computer science (handwritten digit recognition, speech processing [18]) and many other daily life applications [19] can be cited as examples of ANN usage. Software implementations of ANNs, however, do not utilize the parallelism of processing unless special purpose multiprocessor systems are used [20]. As it happens, "simulation" of ANN models may require long training and/or execution times. The term "long" may designate periods of several minutes to several weeks on mainframe computers depending on the complexity of the problem, which can be described by several factors as,

- number of independent variables (inputs);
- number of training samples;
- difficulty of the task to be learned.

On the other hand, the execution time (computation of outputs for a given, trained network) can also be considered as being long depending on the need: In a particle detector system or a fingerprint recognition system, classification of signals has to be achieved in real time, that is, a software implementation using a dedicated microprocessor would not be fast enough [21]. Use of dedicated hardware for ANN has also emerged as a solution offering the following advantages over simulation by general purpose computers:

- independence from a host computer;

- faster learning and forward operation in real time;

- easier interfacing with the outside world in case of analog implementations;

where several disadvantages also exist for dedicated ANN hardware implementations that adversely affect their use:

- necessity of large boards/chips for practical ANN applications;

- nonidealities and mismatches in ANN hardware (especially in analog implementations) and limited resolution of building blocks;

- difficulty in training due to the nonidealities and the probable necessity of a host computer for training;

- storage of weights (digital storage requires memory, analog storage has programming and decay problems [22]).

There is also an important decision to be made by choosing analog or digital hardware: the compactness and speed of analog systems versus the predictable accuracy and ease of programming of digital systems. Despite their advantages, special-purpose ANN hardware is not yet as commercialized as general purpose processors, thus they cannot compete with the advances in the technology and architecture of general purpose computers. As digital hardware for ANN is usually specific for an algorithm, the designer cannot utilize the flexibility of software available on a general computer [23]. Hardware realizations of ANNs are, however, still very promising, since they can exceed computational capabilities of software and advances in VLSI (Very Large Scale Integration) technology may offer low-power, high-density ANN chips operating at higher frequency so that analog ANN blocks can be integrated to a System-On-a-Chip if the following problems are solved:

- circuit based problems (circuit nonidealities, mismatches between identical chips, difficulty of full-custom analog design, weight storage and update, interfacing with other computational resources);

- training based problems (on-chip (use of extra hardware) or chip-in-the-loop (use of extra software) training, limitations on the algorithms due to circuit nonidealities).

It can also be said that the problems mentioned above cannot be considered separately: The training algorithms have to be modified so as to incorporate circuit-based features, whereas the circuits have to be designed such that they allow a meaningful training to be done.

In the literature, analog neural network implementations have been rather ad hoc in that very few, if any, have explored the constraints that circuit nonidealities bring about. Examples are nonlinear synapses [24–25], neurons that deviate from ideal functions, or errors and limitations in storing weights [26–28]. It has previously been shown that multiplier nonlinearity can be a very severe problem even for nonlinearity factors of less than 10% for many applications [24–25], [29–31]. Limited precision in storing weights has also proven to be a crucial problem in analog neural network design. The work in this area has been mostly limited to predicting these effects either through simulation or through theoretical analysis and developing some methods to overcome these problems partially. In [25], the effects of some nonidealities have been studied through circuit simulation with SPICE and the importance of circuit-level simulation in analog neural network design has been demonstrated. Research has been done about the constraints on analog ANN training posed by circuit nonidealities [27, 29], [32–34] and mismatches [34], and strategies have been formulated for the training of hardware by software. In this method, nonideal behavior of analog neural network circuitry is modeled based on simulations or measurements, and the training software employs those models in the backpropagation learning [27, 29, 32, 33]. In fact, this type of modeling and training is usually required for obtaining a suitable initial weight set to be used in chip-in-the-loop or on-chip training. Unfortunately, a perfect modeling is not possible. The models are based either on the results of SPICE simulations or actual measurements that usually can not be expressed analytically. Small deviations from the actual physical behavior in the models may cause the analog neural network training to fail: The training is performed satisfactorily on the software; however, outputs of the actual circuitry deviate heavily from the ideal training set [25]. As a remedy, the authors have, in one of their previous studies, offered ANNSyS (An Analog Neural Network Synthesis System) through which the on-chip training of analog neural networks by software was enabled for backpropagation learning on multilayer perceptron structures. The main idea of ANNSyS can be summarized as follows: Models of the building blocks, namely the synapse (a multiplier), opamp and the sigmoid block, are obtained from their SPICE simulation outputs. An initial backpropagation training based on the models is carried out. Then, the actual circuitry is simulated in a special, fast DC simulator realizing the MADALINE Rule III. Hence, the weights obtained through approximate models are fine tuned by simulating the whole circuit using the most accurate models available (SPICE models), which results in a heavy computational load. Details of this approach can be found in [35]. This method has been shown to be effective in the training of the analog neural network circuitry in software.

An essential requirement for SPICE simulations to have meaning is that the outputs of actual integrated circuits match the simulation results within some tolerance. Furthermore, the outputs of identical blocks in the same chip and on all different chips have to match in order to perform a meaningful design based on those components. Any deviation between the outputs of identical blocks would cause severe adverse effects in the overall system since they are not included in any part of the modeling. Unfortunately, statistical variations in the production process of the integrated circuits result in the deviations mentioned above. Such variations impose serious constraints on the training algorithm of the analog neural network: Each chip has to be trained individually since there are variations among "identical" chips. This makes it impossible to replace a trained chip by an off-the-shelf component. The fault tolerance of the analog neural network hardware is severely degraded by such variations. There are several attempts to enhance the fault tolerance of ANN circuitry: Injecting synaptic noise during training has been shown to improve the fault tolerance performance of ANNs considerably [27, 33, 36]. This could also be used as a means to overcome the problem associated with the training of the ANNs since we can hope that the ANN trained by random noise injection would also be robust against statistical variations at neural network blocks. It has been shown in [34] that noise injection based on modeling the transistor mismatch, and hence having a certain distribution related to the actual hardware, would be more beneficial to noise injection with a random mechanism. Despite the promising efforts in the area of hardware training by software, a general methodology for system-level synthesis is still missing. The usual practice follows constructing building blocks, characterizing them by simulation or measurement, and using the results in the training. This section describes the realization of analog ANNs in VLSI by developing a system-level synthesis tool that will specify the requirements for analog ANN building blocks based on the system-level specifications and constraints supplied by the designer. By applying a simulation based approach, constraints on the hardware models of building blocks will be derived.

The suggested new paradigm of *system level hardware learning by software* will be based on models of analog building blocks. The methodology incorporates hardware-related nonidealities and process-dependent variations into account so that a robust training of analog ANN for a given problem can be achieved solely on software during the design phase of the network where performance requirements of individual blocks will be the output passed to the next level of automation, namely to the circuit-level synthesis tool. The main emphasis will be on feedforward, multilayer perceptron (MLP) type networks realized in MOS technology and trained by the backpropagation algorithm. The main building blocks of the MLP neural network under consideration are,

- voltage-input, current-output Gilbert multipliers to be used as synapses;

- opamps for current-to-voltage conversion;

- sigmoid block as the nonlinear activation function;

as shown in Figure 2.15.

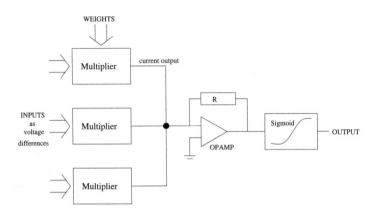

FIGURE 2.15
ANN unit cell.

2.5.1 System-Level Modeling

The system-level synthesis tool will accept the following parameters as input for training:

- structure of the MLP: number of inputs (n_i), hidden units (n_h), outputs (n_o);

- training set (vector of input and desired output tuples);

- backpropagation training parameters: desired error level for convergence, learning rate, momentum, weight decay constant, maximum iteration limit.

- parameters related to the building blocks: maximum allowable input and weight values, supply voltages, slope of synapses, value of R in the current to voltage conversion etc.

Then, the tool will cycle through five loops as shown in Figure 2.16. The idea is to sweep the degree of nonlinearity (given as a percentage for the synapse

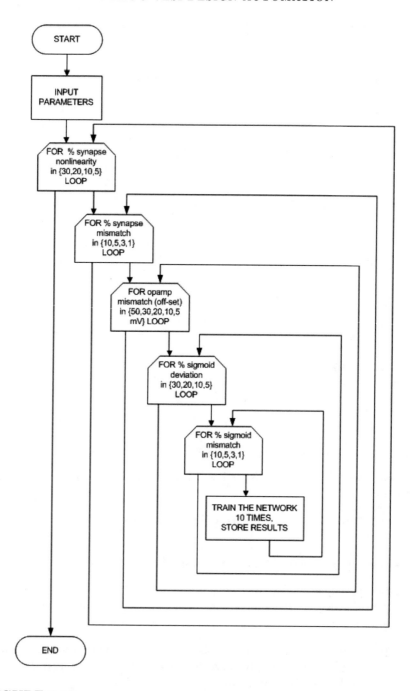

FIGURE 2.16
Flow diagram for system-level synthesis.

and the sigmoid) and random mismatch (given as a percentage for the synapse and sigmoid, and given as the maximum value of the opamps off-set voltage) for several values independently. There are a total of $4X4X5X4X4 = 1280$ combinations! Moreover, 10 independent training cycles are carried out for each combination so that a statistically significant result can be obtained. Hence, a total of 12800 simulations have to be carried out for a single problem to be solved, in order to check all possible combinations. As the process continues, however, some modifications are possible so as to reduce the number of training phases: If all 10 trials for the training of a certain set of deviations converge, then, it can be assumed that the system will also converge for smaller values of mismatch, hence, there is no need for further training. This can simply be stated as follows: If a certain level of nonlinearity and mismatch can be tolerated by the system, then a lower level of mismatch should also be tolerated. Some evidence about this issue will be mentioned later when examples and training performance are discussed. During a single training cycle, all synapse and sigmoid blocks will have the same nonlinearity within themselves. The amount of percentage mismatch to be applied at the building blocks' outputs is a random value that has a normal distribution with zero mean and the deviation (i.e., standard deviation) given by the percentage mismatch value mentioned above. At the start of each single training cycle, the mismatch characteristics of each block is determined: a random number is generated from the corresponding normal distribution and the result is used as additive noise at the output of the corresponding block only. The mismatch characteristics of any block will not be changed during the training phase.

The synthesis tool has to model the ANN building blocks where the following criteria are met:

- the model has to be differentiable so that backpropagation training can be carried out;

- the model should resemble actual ANN building block characteristics;

- the model should be simple for reducing the computational load and there should be an easy way to relate the model with the per cent nonlinearity values.

For this purpose, the synapse function is modeled as given in Equation 2.10.

$$\mu(x, w) = c_0 + c_1 xw + c_2 x^3 w + c_3 xw^3 \tag{2.10}$$

where $\mu(x, w)$ is the output current for the input pair x, w. In the ideal case, $c_1 = constant$ and $c_0 = c_2 = c_3 = 0$.

The user will supply the maximum allowable values for the input and the weight, $| x_{max} |$, $| w_{max} |$, which depend on the supply voltage and the circuit architecture. Moreover, the user will also specify the slope of the synapse, c_1 and the nonlinearity factor for the synapse, Δ_{max} which is the relative deviation between the actual and ideal synapse outputs at maximum input-weight combination as given by,

$$\Delta_{max} = \frac{\mu_{ideal}(x_{max}, w_{max}) - \mu(x_{max}, w_{max})}{\mu_{ideal}(x_{max}, w_{max})} \qquad (2.11)$$

In order to model the nonlinearity, a further assumption is made that the deviation from the ideal case is half the maximum deviation when the weight value is half of the maximum weight (this seemingly magical assumption is based on experience gained in previous designs: Due to the structure of the synapse, the nonlinearity of the output is higher with respect to the input x whereas the output depends with a higher degree of linearity on the weight, w:

$$\frac{\mu_{ideal}(x_{max}, w_{max}/2) - \mu(x_{max}, w_{max}/2)}{\mu_{ideal}(x_{max}, w_{max}/2)} = \frac{\Delta_{max}}{2} \qquad (2.12)$$

Then, the tool computes the nonlinearity coefficients by solving Equations 2.11 and 2.12 simultaneously where we assume that $x_{max} = w_{max} = V_{max}$. The solution yields:

$$c_3 = 2c_2 = -\frac{2\Delta_{max}c_1}{3V_{max}^2} \qquad (2.13)$$

The above mentioned situation is depicted in Figure 2.17 where the synapse characteristics is based on the example design which will be explained later.

In a similar manner, the activation block is analyzed for deviation from the ideal case. The activation function is taken to be a shifted sigmoid:

$$\phi(x) = d_0 + \frac{d_1}{1 + exp(d_2 x)} \qquad (2.14)$$

where in the ideal case $d_0 = -1$ and $d_1 = 2$.

The user will supply the maximum allowable value for the input, $\mid x_{max} \mid$, and the maximum deviation, Δ_ϕ, between the actual and ideal sigmoid outputs at maximum input as given by,

$$\Delta_\phi = \frac{\phi_{model}(x_{max}) - \phi(x_{max})}{\phi_{ideal}(x_{max})} \qquad (2.15)$$

The model will be based on the assumption that the output is symmetrical with respect to the origin. Then, $d_0 \approx -1$, $d_1 = 2 + \Delta_\phi$, and $d_3 = \frac{\ln(0.01)}{\lceil x_{max} \rceil}$ so that the *sigmoidal* function saturates at a maximum input of $\mid x_{max} \mid$ by reaching 99% of the peak value $(d_1 - d_0)$ as shown in Figure 2.18.

Finally, the opamp (I-V conversion) block can be represented by

$$O(\mu) = R\mu + \Delta V_{\text{offset}} \qquad (2.16)$$

where μ is the total current summed at the outputs of synapses, R is the resistance, and ΔV_{offset} is the off-set voltage at the output of the opamp due to mismatches.

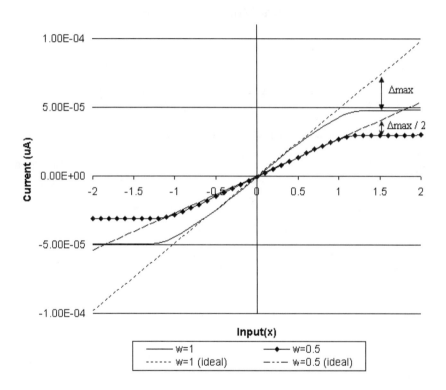

FIGURE 2.17
The synapse characteristics and the nonlinearity.

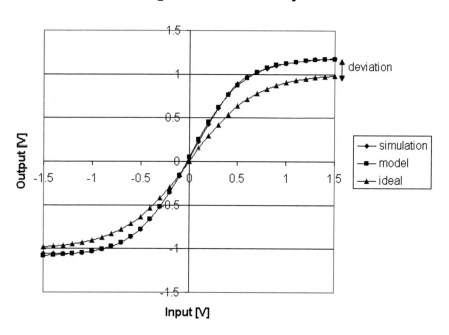

FIGURE 2.18
The sigmoid characteristics and the deviation.

2.5.2 Numerical Examples

In order to verify the functionality of the ANN synthesis system, the examples displayed in Table 2.5 have been studied. Data for the 2-D Classification

TABLE 2.5 Examples for the ANN Synthesis System.

Problem	Inputs	Hidden units	Outputs	Training samples
XOR	2	3	1	4
3-bit Parity	3	6	1	8
2-D Classification	2	14	2	200

problem can be seen in Figure 2.19, where two sets of data points are generated from a normal distribution with the following properties: The one labeled by *class-1* has mean of -0.5 for $x1$ coordinate, mean of zero for $x2$ coordinate, where the standard deviation for both coordinates is 0.3. *class-2*, on the other hand, has mean of 2.0 for $x1$ coordinate, mean of zero for $x2$ coordinate, where the standard deviation for both coordinates is 1.0. Of the 100 data points generated for each class, half are used as the training set, where the other half is used for the test. A 2:14:2 structure is used for training where *class-1* and *class-2* outputs are designated by the pairs of (max-output, min-output) and (min-output, max-output) of the sigmoid as target values. For XOR and 3-bit parity problems, high and low inputs correspond to 1 and 0 respectively. Target values for output are max-output and min-output of the sigmoid. Training was carried out until the rms training error dropped below 1% for all examples.

2.5.3 Results and Conclusion

XOR Problem:

For the XOR case, neither synapse nonlinearity nor sigmoid nonlinearity posed any problems in the training. This is an expected result for the sigmoid since any continuous, non-decreasing function can be used as the activation function theoretically. On the other hand, as long as the nonlinearity can be modeled using a differentiable function on the synapses, backpropagation algorithm also converges. This can be explained by the fact that there may exist local minima on the error surface for the given model of the synapses. This situation is also reinforced by the existence of weight decay in the training algorithm: the weights are pushed toward smaller magnitude so that synapses operate in the *linear* range. Regarding the mismatches, which are reflected as additive noise, the results of training are displayed in Table 2.6, where only the combinations that have yielded successful training are given (i.e., all of the 10 training trials succeeded). A " * " in a row indicates that for all values

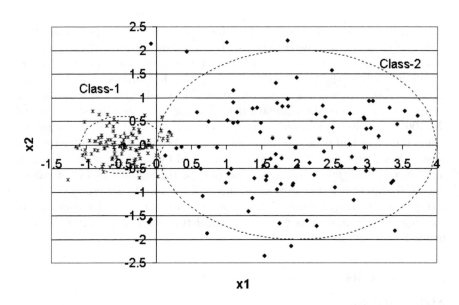

FIGURE 2.19
Data for the classification problem.

in the loop range of the parameter under consideration, the training has been achieved satisfactorily.

TABLE 2.6 XOR Results.

% mismatch in synapse	opamp offset(mV)	% mismatch in sigmoid
*	20	5,3,1
*	10,5	*

3-Bit Parity Problem:

For the 3-bit parity case, sigmoid nonlinearity did not pose any problems in the training. However, 30% synapse nonlinearity caused the training to fail even for cases when there is no mismatch in other components. The combinations of mismatches for successful training are displayed in Table 2.7.

TABLE 2.7 3-Bit Parity Results.

% mismatch in synapse	opamp offset(mV)	% mismatch in sigmoid
5,3,1	20	5,3,1
*	10,5	*

2-D Classification Problem:

For the classification problem, sigmoid nonlinearity did not pose any problems in the training. However, 30% synapse nonlinearity caused the training to yield very low success rates when there is no mismatch in other components. Some combinations of mismatches and the corresponding average success rates in the training are shown in Table 2.8. Based on results, it can be said that the mismatches are more dominant on network performance in comparison with nonidealities. Due to their probabilistic nature, process-based mismatches can not be modeled fully at block level. This makes the issue more critical. For satisfactory results, great attention must be paid to the circuit level design of components so as to minimize effects of mismatches between identical blocks.

Regarding the CPU load, Table 2.9 contains data about the training time required for the above mentioned examples on a PC with 1 GHz Pentium-III CPU and 256 MBytes of RAM. As can be seen, the execution time for large problems can be very long, since an exhaustive sweep is carried out.

TABLE 2.8 2-D Classification Results.

synapse nonlinearity	% mismatch in synapse	opamp offset(mV)	% mismatch in sigmoid	success rate (%)
20	10	20	10	48
20	10	5	10	79
20	10	5	5	83
20	5	5	1	87
20	1	5	1	85
10	10	20	10	46
10	10	5	10	83
10	10	5	5	88
10	5	10	5	84
10	5	5	1	91
10	1	5	1	93
5	10	20	10	57
5	10	10	10	65
5	10	10	5	81
5	5	5	1	94
5	1	5	1	92

TABLE 2.9 Training Time.

Problem	Time
XOR	64 min.
3-bit Parity	5.8 h.
2-D Classification	93 h.

This can be avoided if a certain level of success is accepted as satisfactory: Once that level of success is achieved within a cycle, the training does not need to be continued for subsequent lower levels of mismatch values since training using them would also yield equal or better performance values. The careful reader will note at this point that lower mismatch values translate to larger devices and hence to larger circuit areas. Similarly, higher linearity in the synapses (which will make the error surface smoother and hence training easier) will correspond to more complicated structures, again making area and power consumption higher. The results obtained from training are evaluated through the performance estimator and the optimum design is chosen. A more detailed example run will be presented in Chapter 5.

2.6   High Level Synthesis of A/D Converters

Although modern signal processing is performed mostly in the digital domain, the signals themselves arise from the real world, which is analog. This creates an ever-increasing pressure on A/D converter design, in that faster and more accurate converters are required all the time. The time-to-market pressures and increasing performance demands necessitate design automation tools for A/D converters. This has also prompted many researchers to develop various tools for A/D converter design automation. Some of these tools are similar to the A/D converter synthesis tool described here because they are also top-down and constraint driven approaches [37,38]. However, these efforts and most others present the design of a specific topology. Furthermore, they are not well-defined tools that can be plugged into a general purpose analog design automation system.

There are several different approaches to A/D converter design automation in the literature. One approach is the use of circuit-level simulators. However, traditional circuit-level simulators cannot handle the complexity of most data converter circuits. For circuit-level simulation, all subcircuits of the converter, both analog and digital, are modeled at the device level as an interconnection of basic devices, and the resulting architecture is simulated with a circuit-level simulator like SPICE. However, the complexity of the present converter architectures is beyond the limits of these simulators. For example, a 10-bit flash ADC contains 1024 comparators, each of which can consist of more than ten transistors. If no simple expression for the integral nonlinearity (INL) is available, one general way to verify the nonlinearity specification is to simulate the entire INL characteristic of this converter. This requires many transient simulations, each on a circuit with more than 10^4 transistors. For oversampling converters, high oversampling ratio causes very long transient simulations. Hence, architecture selection or even verification of a single topology consumes an unreasonable amount of time. Alternative approaches have been proposed toward the modeling of different subblocks. These alternative approaches basically aim to trade off accuracy with simulation speed.

A solution for the simulation speed problem can be the use of macromodels for subcircuits such as operational amplifiers and comparators. This approach is useful, but is, in general, limited in that it is hard to find an appropriate equivalent topology for each circuit architecture that accurately matches the behavior of the original circuit, including all nonidealities. Furthermore, the macromodels in most cases do not give hints to the designer about design procedures and tradeoffs.

Another and probably the most popular method is the utilization of behavioral models. A behavioral model directly describes the performance behavior of each analog and digital subblock. For analog circuits, this must not only include the nominal input-output behavior but also the nonidealities. The model can be implemented with different approaches and it can be developed specifically for a particular circuit design, for a particular circuit architecture or, generically, for a whole class of circuits. A first approach consists of simulating the performance of a subcircuit with a circuit simulator and storing the resulting data in a look-up table, as is used in the ZSIM simulator for delta-sigma converters [39]. This is very efficient for final verification after completion of the converter design, but is inefficient during the design phase, as the model is extracted for one particular subcircuit design. Changing the architecture or even changing some parameters in the same architecture requires the complete generation of a new model, and parameterizing the look-up tables is extremely expensive in both generation time and memory storage.

A second approach consists of describing the subcircuit performance as a mathematical subroutine using a standard programming language [40–42]. However, this method is limited with the ability of the representing the formulae explicitly. In other words, the input–output behavior of the ADC should be modeled precisely. On the other hand, this approach reduces the CPU time considerably. Behavioral modeling, with the increasing capabilities of modeling languages such as VHDL-A or VHDL-AMS, is invoking more and more interest in the research community. The advantage of constructing architectures with predefined behavioral models in an HDL is very frequent topic in recent research [43].

Together with the above simulation methods, synthesis systems for specific topologies that are mostly university-based are presented. Although these systems are restricted to a limited architecture set, several successful automatically designed and fabricated data converters are available, some of which (that have been designed with MIDAS [40]) are commercially available.

Notable earlier work in this area can be found in [44]. Other recent synthesis tools are CADIS [45], HiFaDiCC [46], CATALYST [47], OASYS [48]. The capabilities of these tools are given in two excellent reviews [38], [49], which are summarized in Table 2.10. Also, there are other efforts for algorithm-driven synthesis [50] where the architecture is modeled by means of a signal flow graph and a pattern recognition process is executed that maps the recognized graphs to library blocks.

The A/D converter synthesis tool described in this book first selects the proper architecture by using simplified behavioral models, which include non-idealities such as offset, gain error, or nonlinearity. Since the tool should explore more than one architecture space, the models should be as simple as possible in order to achieve reasonable solution in reasonable CPU time. It first selects the architecture by exploring the solution space according to the given constraints. Currently, two topologies, namely flash and pipelined converters, have been modeled and added to the database. Each topology takes the inputs from the user and estimates the performance of the ADC, which results in the generation of subblock specifications such as opamp gain, comparator offset, resistor values, speed of the combinational logic. The power

TABLE 2.10 Previous Work on ADC Automation.

Tool	CADIS	HiFaDiCC	Azteca	CATALYST	OASYS	MIDAS
Archi-tecture	succ. approx. ADC, DAC	succ. approx. & pipelined	succ. approx. & MDAC	flash, subranging, pipelined ADC	pipelined	$\Sigma\Delta$ ADC
Technology	CMOS	CMOS	CMOS	CMOS	CMOS	CMOS
Optimi-zation	yes	yes	yes	yes	yes	yes
Layout	UCB layout tools	yes	no	no	no	Philips layout tools
Verification	behavior sim.	general sim.	behavior sim.	behavior sim.	dedicated formulae	behavior sim.
Working Si	yes	not reported	not reported	not reported	not reported	yes

of this methodology is that it is not limited with the available library. If no solution is available after the solution space or the library database has been searched, then the methodology widens the solution space, which is only limited by the parameters given by the user. The solution can be found by sweeping these parameters, which are total power, total area, speed, comparator offset, and area for the flash converter. The methodology developed generates a solution set for each topology, which consists of different ranges of requested parameters. Also, the performance estimator can be used in conjunction with the synthesizer to choose the optimum solution in a given solution set. Another advantage of this methodology is that the user can limit the solution space in any dimension. In other words, the user gives the ranges of all parameters. The tool decomposes the problem at hand in the manner shown in Figure 2.20. During the sweep operation, the performance estimator will be employed to calculate the costs associated with the circuit blocks that are to be designed during the circuit-level synthesis.

The fastest of all types of high-speed analog-to-digital converters is the flash or parallel type of converter. The resolution of flash converters tends to be limited to eight bits in practice due to the fact that the amount of circuitry doubles every time the resolution is increased by one bit. A block diagram of a flash converter is shown in Figure 2.21. Even though the design of flash converters is highly repetitive, it demands a high level of matching between the parallel comparators. One of the major contributors to the nonlinearity is the comparator input offset voltage. The offset should be less than $\pm 1/2$ LSB not to degrade the monotonicity of the converter. Similar effects could be caused by the bias and input offset currents of the comparators. Together with the resistance of the reference ladder, they will be added to the offset voltage. The reference voltage resistive ladder contributes as a secondary effect to the error. The error sources and their effects are discussed below:

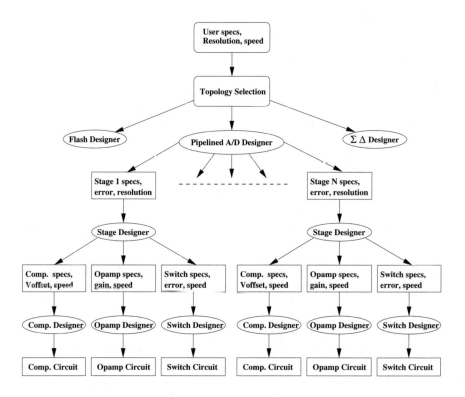

FIGURE 2.20
ADC decomposition example.

Power of the resistor string is calculated as follows:

$$P_{res} = \frac{V_{ref}^2}{2^n R} \tag{2.17}$$

where P_{res} represents the power dissipated at the resistor string. The powers dissipated at the other partitions of the ADC are:

$$P_{comp} = (2^n - 1)(I_{sup})(V_{sup})$$
$$P_{dig} = C_{dig} V_{dd}^2 f \ k \tag{2.18}$$

where, P_{comp} and P_{dig} are powers of comparators and digital circuitry. The f value represents the frequency of the encoder and k is the switching probability of the encoder. C_{dig} is the overall capacitance of the encoder.

Like the power estimation, the area estimation of the resistor string also requires technology parameters. From the resistance values, the ratio of W and L are calculated. Since the length of the resistor is indirectly proportional to the resistance and the smaller resistance area is desired to decrease total area, the length has been set to the minimal value. The minimum allowed

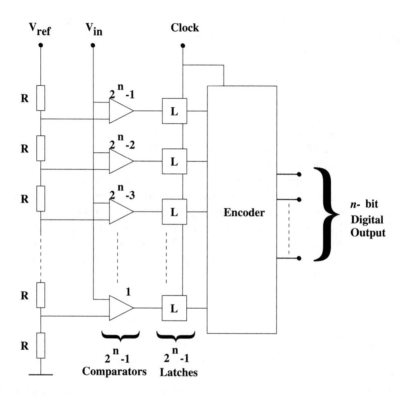

FIGURE 2.21
Flash ADC.

length for the technology used in the examples is 0.7μm. Polysilicon is used to implement the resistances. The sheet resistance of polysilicon, R_s, is read from the technology file. The resistance can be calculated as follows:

$$R = \frac{w}{l}R_s \qquad (2.19)$$

The resistance can be one of the desired outputs. In that case, the width value is calculated from the given resistance. The area of the resistor string is simply:

$$A_{res} = w \, l \, 2^n \qquad (2.20)$$

The area of the comparator is an input parameter. The total area of the comparators can be calculated as follows:

$$A_{comp} = (2^n - 1)a_{opamp} \tag{2.21}$$

The area of the opamp is critical for the total area since for increasing n, A_{comp} increases exponentially.

The third stage is the digital part, in other words, an n-bit encoder. The parameters of the encoder, like number of gates for n-bit encoder should be available. The methodology uses a parameter file for encoder block. The parameter file is a look-up table generated for n bits. A generic C code was written in order to generate the VHDL code of the n-bit encoder. Then, the VHDL code was synthesized. Also, available optimizations were applied and the number of gates for each n were written to the parameter file. The synthesized designs were then simulated with a digital simulator and the observed delay values were also entered into the parameter file.

The area of the digital circuitry is directly proportional to the number of gates used. The area of each gate can be found from a technology data book. The areas are specified in terms of uca (unit cell area). Since different synthesizers may use a different number of gates and structure, the area is calculated with an average gate area. The minimum and maximum gate areas must be specified in the technology file. The number of D flip-flops can be easily calculated, so the area of D flip-flop is calculated separately. The area of the digital circuitry is as follows:

$$A_{dig} = k_2 \left[((n + 2^n - 1)a_{dff}) + \left(\frac{a_{min} + a_{max}}{2} n_{gates} \right) \right] uca \tag{2.22}$$

The total area of gates cannot represent the total area of the circuitry. The area used for routing should be added. The coefficient k_2 is used for this purpose.

The first stage of the flash ADC is a resistor string. In order to show the thermal noise in the resistor string, a voltage source is added to the string for each resistor. The thermal noise generates a voltage e_t

$$e_t = \sqrt{4KTR} \times BW_{comp} \tag{2.23}$$

Also, the mismatches of the resistors contribute another voltage source. The mismatch is the standard deviation of the relative differences of two identical resistors. The models of the resistors, which are at most $10\mu m$ from each other, are used in the algorithms to calculate the mismatch. For more accurate estimation, the resistors should not exceed this range. The resistance values are functions of w, l and the sheet resistance (R_s). The maximum deviation from the standard value can be calculated by using the 3 sigma design method. As a result, mismatch contributes a voltage difference:

$$e_m = \frac{V_{ref}}{2^n R} 0.03 \ m \ R \tag{2.24}$$

where m is a mismatch parameter that is read from the technology file. The sum of these voltages with the offset voltage of the comparator should not exceed half of the voltage, which is equal to the one LSB.

$$e_m + e_t + v_{os} < \frac{V_{ref}}{2^{n-1}} \qquad (2.25)$$

The flash architecture is fast enough to give the digital output in one clock cycle. Hence, the delay for this architecture is one clock cycle. The clock frequency is the only factor in speed estimation. The major factor in defining the clock frequency is the speed of the comparator. The comparator, mostly a clocked one that includes a latch at the output, should be fast enough to operate in several MHz range. Various comparator designs are available in the literature. The most important specification of the comparator is the slew-rate. In the methodology, the slew-rate is calculated from the rise or fall time. Half a period is reserved for the latch and the digital circuitry. Hence, the comparator should give the correct output at the other half of the clock period. One other factor that should be taken into account is the minimum and maximum voltages for the logical "1" and "0." Since the output of the comparator exceeds the logical "0" or logical "1" threshold, the latch can fetch the right logical value. There is no need to wait until the comparator output reaches the maximum or minimum value.

The module for flash converter operates in the same order as presented above. First of all, the desired SNR should be given to estimate the effective number of bits (ENOB) for the topology. The SNR is limited from the bottom by the quantization noise, which means that the quantization noise determines the minimum number of bits for the desired SNR. The expression below shows the mathematical expression for the quantization noise [53]. The n represents the number of bits required to achieve the desired SNR.

$$SNR = n \times 6.02 + 1.76 \ \ dB \qquad (2.26)$$

The given SNR limits the number of bits. The methodology proposed for flash converter in this book is limited by eight bits. This is mainly because the pipeline architecture offers better results by means of complexity, area and power. Also, knowing the least number of bits limits the solution space considerably.

The second step of the methodology is to read the technology parameters. Since these parameters do not change while the functions of the topologies are executed, they should be read only once per each program run. Also, the other parameters used by the methodology are read from the files. For example, the parameters for the comparator are read at the beginning and stored at the memory for fast processing. However, this may not be efficient for a large library database.

After the parameters are read, the power of the available solutions is generated. This solution is limited by the maximum power given by the user.

Area is the next step in the methodology. Since power estimation reduces the number of solutions in the design space, area estimation deals with a narrowed design space. Area estimation again decreases the number solutions. At this time, since more precise values are determined by the previous stages, errors in the flash architecture are calculated in order to eliminate the designs that will not function appropriately.

The last step is to estimate the speed of the architecture and check whether the solutions fit the given specifications. After the speed estimation, the solution space is available. However, the optimum solution should be determined by means of area, power and speed. The optimum solution can be achieved by a cost function, which is the case in this methodology. As the functions are reducing the number of solutions, a cost function also processes the solutions in order to determine the optimum one. The optimum solution can be adjusted by the parameters given by the user.

A pipeline ADC consists of a number of consecutive stages. The stages are similar in their function and each stage generally resolves only one or two bits. Each individual stage consists of a sample and hold, a low resolution flash ADC, a low resolution digital-to-analog converter (DAC) and a summing stage including an interstage amplifier for providing gain. The outputs of each stage are combined in the output latch. Stage one takes a sample of the input voltage and makes the first coarse conversion. The result is then the most significant bit (MSB) and its digital value is fed to the first latch. As the residue of the first stage gets resolved in the subsequent n-stages, the MSB value is rippled through n latches in order to coincide with the end of the conversion of the last stage. Then, all data bits are latched in the output register and the output becomes available.

Figure 2.22 shows a typical pipeline ADC architecture. The structure is highly repetitive. There are NS identical stages, each quantizing N_i bits. Thus, the overall resolution is NS times N_i. Each stage samples the output from the previous stage and quantizes to N_i-bit digital codes using flash ADC architecture. The codes are then converted to an analog signal by N_i- bit DAC and subtracted from the sampled signal. Next, the residue is amplified by a gain of $G = 2^{N_i}$. The output register combines the output bits from each stage and gives the final digital codes. Stage one gives MSB's, stage NS gives the LSB's.

Due to the small dimensions and low power consumption, the pipeline architecture is more suitable for high-resolution applications than flash converters, but is also easily affected by circuit imperfections. The errors caused by these imperfections are presented below. Although many error sources are available, correction circuitry can compensate for these errors efficiently.

The pipeline architecture offers relatively high speeds and approximately linear hardware cost with resolution. A very important advantage of pipeline architecture is that it allows digital error correction and gain calibration (for offset/gain errors, and nonlinearities) and power minimization through capacitor scaling. Pipeline ADC is most popular in high speed (several MHz-

50MHz) and high resolution (above eight bit) applications where latency is not a concern.

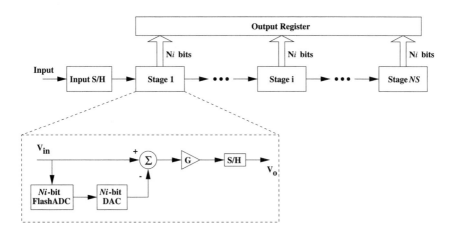

FIGURE 2.22
General model of a pipeline ADC.

The main advantage of pipeline ADC's is that they can provide a high throughput rate with moderate design complexity and low power consumption. This is because of the concurrent operation of the NS-stages. The associated data latency is not a limitation in most applications. Two main clock phases are required per conversion because the pipeline ADC uses flash converters. Therefore, the maximum throughput rate can be high. After the initial data latency time, the data representing each succeeding sample is output with every following clock pulse.

In recent studies, multiplying DAC's (MDAC) are utilized to carry out the combined functionality of three components; namely, sample and hold (S/H), DAC, and residue amplifier. Figure 2.23 shows the architecture of a recycling ADC with MDAC's. Recycling converters can easily be transformed into pipelined converters by feeding the residue to the next stage instead of recycling. The architecture is very straightforward since every MDAC already has its own built-in S/H.

Below, the errors in a general pipeline architecture are presented. The methodology proposed uses an MDAC, which combines and eliminates some of these errors. The dominant error sources of an MDAC are also presented below. The main error sources in a general pipeline architecture are gain, offset and nonlinearity errors in sub-ADC (SADC), DAC and sample and hold amplifiers (SHA). The other errors such as settling errors can be modeled as a combination of these errors [51]. A model for errors in a pipeline ADC has

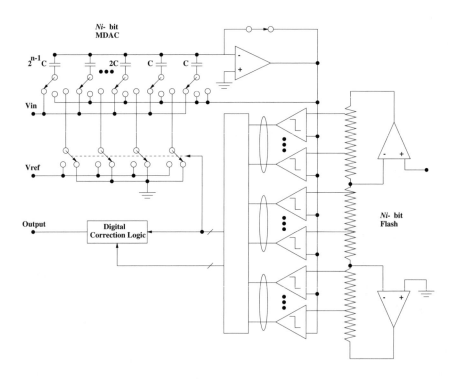

FIGURE 2.23
The recycling two step converter [54].

been introduced in [51]. Since then, many researchers have worked with the same model [52–54].

The model of an analog path in a pipelined ADC is presented below. It consists of NS stages, each containing an SHA and an error source, e_i. In general, each e_i represents the offset, gain, nonlinearity, and quantization errors in that stage. The input-referred error, e_{in}, which is equivalent to the contributions of all the individual error sources is,

$$e_{in} = e_1 + \sum_{i=1}^{NS-1} \frac{e_i}{G_i} \tag{2.27}$$

where G_i stands for the the product of gains (G) of the stages before the stage i. The general expression for the gain is as follows:

$$G = 2^{Ni-x} \tag{2.28}$$

In this equation, x represents the number of bits for redundancy. One bit redundancy is used in this methodology, which can handle errors as large as the amplified residue. The previous equation shows that SHA gains greater than one reduce the effects of nonidealities in all stages after the first on the error of the entire conversion. For each nonideality, the reduction factor is the product of all the SHA gains before the nonideality except the SHA gain in the first stage. As a result, to limit the ADC error arising from each error source to less than 1/2 LSB, the equation below should hold:

$$e_i \leq \frac{FS}{2^{n+1}G_{i-1}} \tag{2.29}$$

where FS stands for full-scale conversion range. If errors are similar then,

$$e_i = e, \text{ for } 1 \leq i \leq NS \tag{2.30}$$

In this case, the input referred error can be rewritten as:

$$e_{in} = e \left(1 + \sum_{i=1}^{NS-1} \frac{1}{G_i} \right) = eF \tag{2.31}$$

This equation shows that the combined effect of identical errors in all stages is greater than the effect of only the first-stage error by a factor, F, which depends on the SHA gains. If $G = 1$, $F = NS$, and if $G \gg 1$, $F \sim 1$. A boundary between these two limiting cases occurs when $G = 2$ and $F \sim 2$. Therefore, to limit the effect of errors in all stages after the first so that the first-stage error dominates the total ADC error, the stage resolution should be chosen so that $G \geq 2$. From that, we can conclude large stage resolution reduces the error of multistage ADC's. The errors of each functional block can be modeled in a similar way. An overview of these blocks with the constraints they introduce is presented below.

A sub-ADC is used in the pipeline architecture. A flash quantizer can be preferable among others since 2-bit stage resolution is very common in the pipeline architecture. SADC offset, gain, and nonlinearity errors move the SADC decision levels. The correction range is defined as the amount of SADC decision-level movement that can be tolerated without error in the ADC output code. In all stages except the last, redundancy and digital correction can eliminate the effects on the ADC linearity of all SADC errors up to the correction range. Since the output of the last stage is not corrected, however, SADC errors there do cause ADC nonlinearity, but in an amount that is diminished by the combined interstage gain before the last stage. The stage redundancy x is usually selected to provide enough correction range to eliminate the effects of these SADC errors in all stages except the last.

The main errors of a DAC are offset, gain and nonlinearity. These errors can be modeled as an input referred error. For DAC offsets, the error can be replaced by an input-referred stage offset and an offset in series with the SADC. If the combination of all errors that shift the SADC decision levels does not exceed the correction range, the effect of the SADC offset is eliminated by the digital correction. The remaining input-referred stage offset is equivalent to a DAC offset in the $(i-1)$ st stage. Using this process repetitively, the offsets in all the DAC's can be referred to the input of the ADC. Therefore, to eliminate the effect of DAC offsets on ADC linearity, the stage resolution, n, and redundancy, x, are usually selected so that the correction range is not exceeded by the combination of all errors that shift the SADC decision levels.

DAC gain error can be replaced by three gain errors: one in series with the stage input, one in series with the SADC, and one in series with the stage output. A nonlinearity error in the ith DAC can be modeled by e_{off} that depends on the DAC output. Although the digital outputs of this stage are correct with such an error, the residue output of this stage is incorrect exactly by the amount of the DAC nonlinearity. To limit the resulting nonlinearity to $\pm 1/2$ LSB,

$$e_{offi} \leq \frac{FS}{2^{n+1}} G_{i-1} \qquad (2.32)$$

The first stage again is the dominant source in error. If the stage resolution is selected so that only two-level DAC's are required, the DAC's are inherently linear, and the accuracy of the SHA gain determines the ADC linearity. If the DAC's have more than two output levels, however, they are not inherently linear. To overcome ADC nonlinearity errors caused by component mismatches in such DAC's, self-calibration techniques in pipelined, multistage ADC's with more than 10-bit resolution can be applied. An offset in the ith SHA is equivalent to an offset in the $(i-1)$ st DAC. Therefore, the conclusions made earlier about the effects of DAC offsets apply also for SHA offsets.

The expression of gain error e_{gain} can be reduced to the equation below [51].

$$|e_{gain}| = \frac{G_{i-1}}{2^n} = \frac{1}{2^{n-(Ni-x)}} \qquad (2.33)$$

This equation shows that each SHA gain (after the first) must be accurate enough to preserve the combined linearity of the resolution remaining after the SHA. For $2 \leq i \leq NS$, residue is maximum when $i = 2$. Therefore, the allowable SHA gain error is minimum for the second-stage SHA (which is the first interstage SHA). The equation also shows that increasing n reduces the required gain accuracy because the resolution remaining after the first interstage SHA decreases as N_i increases.

A nonlinearity error in the ith SHA can be modeled similar to the gain error, which is dependent on the residue generated by the previous stage. Because the ADC operates on the SHA outputs, SHA nonlinearity causes an ADC nonlinearity in an amount that depends on the ADC resolution remaining after the SHA. Therefore, to limit the resulting nonlinearity to $\pm 1/2$ LSB, the fractional variation in the gain of the input SHA must be less than one part in 2^{n+1}. In practice, SHA linearity does not usually limit ADC linearity except in ADC's that use correction techniques to overcome interstage gain errors and DAC nonlinearities.

Some of the errors presented above affect the number of stages in a pipeline. But the errors are not the only factor for the optimum stage number and also in stage resolution. For example, contribution of gains in the input referred error expression lead us to the stage resolution of two bits. However, factors like area, power and speed (conversion rate) must be taken into account. In [51] it is shown that the effect of conversion rate is limited by the stage resolution. In order to maximize the conversion rate the minimum stage resolution should be used. Similar to these, the area and power factors add new constraints. The problem of minimizing the total area can be reduced to minimizing the total number of comparators in an ADC, which is done by using the minimum stage resolution. Therefore, the minimum stage resolution minimizes the required die area. By using a similar approach, it can be shown that reducing the stage resolution can minimize the power [51]. As a result, 2-bit stage resolution decreases the area and power and increases the conversion rate for pipelines up to 10 bits.

A common pipeline architecture is given below to show the usage of MDAC in a multistage pipeline ADC.

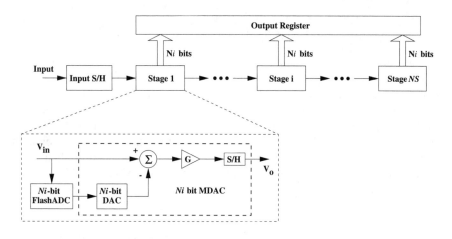

FIGURE 2.24
Pipeline architecture with MDAC.

In Figure 2.24, the MDAC replaces the blocks DAC, SHA and residue amplifier. The structure of MDAC was shown in Figure 2.23. The basic building blocks of an MDAC are an amplifier and a capacitive array. The MDAC has three phases: sample, hold, residue amplification. Because of the capacitive nature of the load (the input capacitance of the SADC and the input capacitance of the MDAC of the next stage), the amplifier employed in the MDAC can be implemented using a single-stage operational transconductance amplifier (OTA).

If we consider modern technologies where chip sizes continue to shrink, the capacitor array of the MDAC introduces a new source of error, thermal noise. Since the capacitance values are getting smaller and smaller, thermal noise becomes comparable to or even dominant with respect to the other error sources.

The maximum resolution achieved with the MDAC introduced above is limited mainly by thermal noise, by the nonlinearities of the MDAC's produced by capacitor mismatches, and also by the residue amplification error that is due to the amplifier nonidealities [52-53]. The thermal noise can be reduced by appropriate sizing of the capacitors and the nonlinearities in the MDAC can be reduced by self-calibration techniques and by an appropriate amplifier design.

Error sources other than MDAC's are SADC's. But the errors from the flash quantizer can be digitally corrected if they are kept within the range covered by the redundancy created between consecutive stages. For this purpose, an extra redundancy bit can be added to the architecture [53], [54].

Thermal noise can be written as

$$TTN = \sqrt{\frac{N^2_{S/H}}{1} + \frac{N^2_{MDAC1}}{G^2_{S1}} + \frac{N^2_{MDAC2}}{G^2_{S1}G^2_{S2}} + \cdots + \frac{N^2_{MDAC(NS-1)}}{G^2_{S1}\cdots G^2_{S(NS-1)}}} \quad (2.34)$$

Considering that the main sources of thermal noise are the on-resistances of the switches and the opamps of the MDAC's and of the front-end S/H (can be implemented by using a MDAC), then the rms value of the total thermal noise (TTN) referred to the input of the converter is given by Equation 2.34, where $N_{S/H}$ and N_{MDACi} respectively, are the output-referred RMS noise contributions of the S/H and MDAC, and G_{Si} represents the closed-loop gain of each stage during residue amplification. According to the simplified noise model shown in [53], the mean square noise contribution of each MDAC, N_{MDACi}, will be a sum of three terms.

$$N^2_{MDACi} = V^2_{S/H} + V^2_1 + V^2_2 \quad (2.35)$$

The first term is a sampled-and-held component introduced by the on-resistance of the CMOS switches during the sampling phase. The second term is a broadband contribution due to the on-resistance of the CMOS switches during the residue amplification phase. The third term is also a broadband contribution introduced by the amplifier itself during residue amplification. Thus, the output-referred mean square noise introduced by each MDAC can be expressed approximately as:

$$N^2_{MDACi} = \phi \left[\frac{KT}{C_{MDACi}} + (4KTR_{ON} + 4KTR_{EQ})\,BW \right] G^2_{Si} \quad (2.36)$$

where C_{MDACi} is the total input capacitance of the MDAC, K is Boltzmann's constant, and T is the absolute temperature. The constant ϕ can take the values of either one or two depending on, respectively, whether the circuit is implemented in a single-ended or fully differential configuration. R_{ON} and R_{EQ}, respectively, are the on-resistance of the CMOS switches during the residue amplification phase and the pseudoresistance at one input of the amplifier caused by switch capacitors [53].

To obtain a pipeline with n effective bits of resolution, one should guarantee that the total thermal noise is below the quantization noise and, thus, for a given reference voltage, the condition [53],

$$TTN < \frac{\phi V_{ref}}{2^n \sqrt{12}} \qquad (2.37)$$

must be satisfied. We can conclude that the capacitor sizes must be carefully chosen to limit the thermal noise contribution.

One of the main sources of the errors is mismatch errors of the capacitor array [55-57]. The errors can cause gain and nonlinearity errors [58]. To show the effect of the mismatch, the ideal amplified residue expression and the expression of the output of the MDAC affected by the mismatches at the capacitor array are presented next.

$$V_{RA} = 2^{Ni-1} \left(v_{in} - V_{ref} \sum_{i=1}^{Ni} 2^{i-Ni-1} b_i \right) \qquad (2.38)$$

The ideal amplified residue of a MDAC is presented in Equation (2.38), where b_i is the binary number converted by the DAC. A feedback capacitor with twice the capacitance of a unit capacitor gives a gain of 2^{Ni-1}. The weighted sum of the capacitors determine the voltage that will be subtracted from the input. If a mismatch, shown by ϵ, is present, the relative capacitor values are,

$$C_f = (1 + \epsilon)C_{unit}, \quad C_i = (2^{i-1} + \epsilon_i)C_{unit}, \quad i = 1, 2, \ldots, N_i \qquad (2.39)$$

With the capacitor values above, an error contributes the expression

$$V_{RA} = 2^{Ni-1} \left(v_{in} \left[1 + \frac{1}{2^{Ni}} \left(\epsilon + \sum_{i=1}^{Ni} \epsilon_i \right) \right] - \frac{V_{ref}}{2^{Ni}} \sum_{i=1}^{Ni} [2^{i-1} + \epsilon_i] b_i \right) \qquad (2.40)$$

By comparing the two error expressions, we can conclude that mismatch error contributes a gain error and another error that depends on the code applied by SADC. The first error term in the above equation represents the gain error. The second error term shows the nonlinearity error of the MDAC. The gain error that affects the input voltage can be referred as an analog error, whereas the nonlinearity error depends on the digital code applied. But the calibration techniques offer many solutions to these errors. The errors in both analog and digital domain can be calibrated with efficient digital correction mechanisms.

The limited gain of the amplifier also generates an error. To limit the effect of this error,

$$A_0 \geq 2^{M-Ni} \left(2^{Ni} + \frac{C_p}{C_{unit}} \right) \tag{2.41}$$

must be satisfied. The M is the accuracy needed by MDAC. C_p represents the parasitic capacitance at the input of the amplifier.

The power estimation for the pipeline can be computed in a straightforward way as:

$$P_{tot} \simeq \sum_{i=0}^{NS-1} P_{MDACi} + \sum_{i=1}^{NS} P_{Flashi} + \sum_{i=1}^{NS} P_{synci} + \sum_{i=1}^{NScal} P_{selfi} \tag{2.42}$$

The functional blocks in pipeline ADC can be classified into four. The blocks are, MDAC, flash SADC, synchronization circuitry and, if available, the self-calibration circuit. However, the first two terms usually dominate the total power.

The power dissipated in each MDAC is the sum of the static power dissipated in the OTA and a dynamic contribution corresponding to switching of the capacitors at the sampling frequency, and can be approximately given by:

$$P_{MDACi} = (I_{supa} V_{supa}) + \phi \, C_{MDACi} \, (1 + K_{Ci}) \, V_{ref}^2 \, f \tag{2.43}$$

where the constant K_{Ci} reflects the parasitic capacitors mainly because of switches.

The power dissipated in the flash can be determined by the static contribution of the preamplifiers of the comparators, by the dynamic contribution, which depends on the energy consumed by the latch per comparison and on the switched capacitances in each comparator, and by the static contribution of the R-string. This is approximately given by:

$$P_{flash} = (I_{sup} V_{sup}) 2^{Ni-1} + C_S \, V_{ref}^2 \, f \, K_F + \frac{V_{ref}^2}{2^{Ni} \, R} \tag{2.44}$$

C_S represents the total switching capacitance in the flash quantizer. The constant K_F represents the probability of switching. R is the unit resistance in the R-string.

$$P_{synci} = N_i(NS + 1 - i) \, P_{ff} \, f \tag{2.45}$$

The P_{FF} is the energy consumed per cycle in a flip-flop.

Also, if available, the self calibration circuitry consumes some power that is directly related to the power consumed in each fetch of the RAM [59], [60].

$$P_{self} \sim P_{RAM} \, f \tag{2.46}$$

The P_{RAM} stands for the power consumed by the RAM in each fetch operation.

The total area can be found by:

$$A_{tot} = \sum_{i=0}^{NS-1} A_{MDACi} + \sum_{i=1}^{NS} A_{flashi} + \sum_{i=1}^{NS} A_{sync} + \sum_{i=1}^{NScal} A_{self} \qquad (2.47)$$

Again area can be classified into four blocks.

$$A_{MDACi} = A_{OTA} + \phi\, A_{Cap.\ Array} + \phi\, A_{switches} \qquad (2.48)$$

The functional blocks of the MDAC are an amplifier (generally an OTA), a capacitor array and the switches that determine the phase of the MDAC.

$$A_{flashi} = A_{res} + A_{comp} + A_{dig} \qquad (2.49)$$

The area of the comparator is the dominant factor in this expression. The detailed expression was given above in the flash converter discussion

$$A_{synci} = N_i\,(NS + 1 - i)\,A_{FF} \qquad (2.50)$$

The area of the synchronizer can be calculated by the number of flip-flops and the area of a flip-flop. The last contribution to the expression is the area of the self-calibration circuitry. The area of the calibration techniques is mostly determined by the area of the RAM, which is used to store calibrating codes. Hence, the area of the calibration circuitry can be reduced to the area of the RAM.

The delay estimation is straightforward. The delay of a pipeline with NS stage is NS times the clock. Actually, the delay is determined by the first stage. The resolved bits in the first stage are delayed until the last stage gives its output. Thus, the synchronization between stages can be obtained.

The speed limitation is determined by the settling time of the amplifier. The same approach, presented earlier, is used in generating the speed requirement. Since the MDAC has different phases, the dominant phase should be calculated. Earlier studies show that the residue amplification phase brings tightened constraints. Hence, the speed is determined by the slew-rate of the amplifier.

Capacitor sizing is basically a low-power technique that reduces the power stage by stage by decreasing the capacitor values, which increases the thermal noise. So, the downsizing of the capacitors must be carefully processed. In [59] it is shown that the two ways to reduce the power are the capacitor scaling and pipeline resolution scaling. The choice of capacitor scaling is implemented in proposed methodology.

Capacitor scaling can also be done as we apply resolution scaling. However, previous work shows that this method actually does not affect the total power. With resolution scaling, the stage resolution of the input stages can be as much as five bits. Hence, the power is mainly consumed by these stages. The downsizing of the latter stages does not reduce the power significantly.

On the other hand, if identical stage resolutions are used, capacitor scaling may decrease power considerably. Earlier studies show that global minimum for capacitor scaling factor is two [60]. This value is used in the methodology where capacitor scaling is requested.

One of the advantages of the pipeline architecture is that it allows calibration and correction. Digital correction is used in order to relax the constraints, mainly on the SADC. On the other hand, self-calibration methods are very efficient and should be used in pipeline architectures over 10-bit resolution.

After getting the first parameter, which is the SNR, the tool asks for some other parameters. These are maximum power, maximum area, maximum resolution, maximum capacitor array size. The design space is limited mainly by the maximum resolution and the minimum resolution calculated from the given SNR. The next step is to generate all available design configurations. Each configuration has different stage resolutions. The design space created by this process should be limited or reduced. Otherwise, reasonable CPU times cannot be achieved. The methodology uses some heuristics to limit the design space. The methodology rejects the configurations, which have low resolutions at the early stages and high resolutions at the latter stages, because low stage resolutions contribute higher thermal noise and the effect of noise at the early stages is crucial. The input-referred noise model shows that the noise at the early stages is dominant. Therefore, the early stages should have higher, or at least equal, stage resolutions.

After the design space is explored and limited by the possible stage resolutions, the thermal noise is calculated and checked. This check may reduce the number of solutions, especially for low reference voltages.

The next step in the methodology is to calculate the required accuracy of each MDAC. Stage resolutions and the maximum matching between the capacitors determine this accuracy. Stage by stage, this accuracy requirement decreases. Hence, the limitation for the residue amplifier loosens. These values are calculated by the methodology for each design and each stage. The result gives the number of stages that require calibration. Also, the accuracy limits determine the minimum gain. The methodology then searches its database to check whether the gain requirements are met. Actually, this step and the following steps basically try to find a library element that does not violate the limits. After this database search, the methodology estimates the power, the area and the speed.

A design example will now be presented. The example is limited by 13 bits. The desired SNR is 75 dB. This ratio determines the minimum resolution. The matching between capacitors is 6 bits. The conversion range is selected as 20MS/s. A reference voltage of 5 volts is used for this example. The number of available configurations is 33 for 12-bit design and 37 for 13-bit design. Some of the configurations are given in Table 2.11.

The configuration column shows the stage resolutions from left to right. The rightmost number shows the resolution of the last stage. The sum of the stage resolutions exceeds the desired resolution because of redundancy used.

TABLE 2.11 Different Configurations for 12 and 13 bits.

Resolution	Configuration	Number of Amplifiers	Number of Comparators	Stages Requiring Calibration
12	22222222222	11	33	6
12	332222222	9	35	4
12	3333222	7	37	3
12	43222222	8	40	3
12	44332	5	47	2
12	54322	5	59	2
13	222222222222	12	36	6
13	333222222	9	39	3
13	3333322	7	41	3
13	443322	6	50	2
13	5443	4	68	2

The methodology uses 1 extra bit for each stage. Thus, digital correction can be performed in each stage.

As seen from Table 2.11, different configurations for stage resolutions may lead to very different designs. The power and area may vary significantly and their values can be calculated either from the performance estimator or by using lookup tables from previous designs. The optimum stage resolution can be determined from a graph similar to the one in Figure 2.25.

The methodology selects the configuration "3333322." Earlier studies about optimum stage resolution [51] conclude that minimum stage resolution is optimum for resolutions up to 10 bits. However, recent studies show that medium resolution stages yield better results above 10-bit ADC resolutions [53]. The results obtained here are similar. This is because high resolutions at the input stages require large number of comparators. The exponential

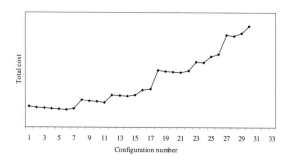

FIGURE 2.25
Optimum stage resolution.

growth of comparator numbers increases the area significantly. That is why configurations like "54322" or "5532" are far from global minimum. On the other hand, low resolution per stage leads to large number of stages. Hence, exponential growth can be observed at the synchronization circuitry. Also, the number of stages that require calibration increases. In conclusion, optimum solution can be achieved by a medium number of stages with medium resolution.

Although the optimum number of stages may be determined, the other constraints have to be verified. These constraints are generated by thermal noise and the error caused by the limited gain of the residue amplifier. Table 2.12 shows the required values for these errors.

TABLE 2.12 The Constraints for the Resolutions.

Resolution	Configuration	Minimum Amplifier Gain for First Three Stages (dB)	Thermal Noise
12	22222222222	78,72,66	2.609×10^{-5}
12	332222222	78,66,54	2.320×10^{-5}
12	43222222	78,60,42	2.211×10^{-5}
12	54322	78,54,36	2.157×10^{-5}
13	222222222222	84,72,66	2.609×10^{-5}
13	333222222	84,66,54	2.319×10^{-5}
13	443322	84,60,42	2.209×10^{-5}
13	5443	84,54,36	2.157×10^{-5}

Table 2.12 shows that the gain requirements for the amplifier decrease through the pipeline. Therefore, the amplifier selected from the library for the first stage may be unsuitable for the later stages. The cost function should take this property into account. This feature was built into the methodology. However, this feature totally depends on the available library elements.

Another feature of the methodology is capacitor scaling. Capacitor scaling results reduce the power dissipated stage by stage. On the other hand, the thermal noise increases with capacitor scaling. However, since the noise generated is reduced by a factor that is the total gain contributed by the early stages, the total thermal noise can be limited below the quantization noise.

Table 2.13 compares the total thermal noise values for different configurations. In the example, the noise stays below the limits determined by quantization noise. This is mostly because of the reference voltage and the resolution. Resolutions about 15 bits may decrease the boundaries significantly. In that case, the total thermal noise constraint may reject most of the configurations.

As it can be seen from the table, the quantization noise does not limit the number of realizable configurations. However, when resolutions of 15 bits are

TABLE 2.13 Effect of Capacitor Sizing on Thermal Noise.

Resolution	Configuration	Total Thermal Noise with Scaling	Total Thermal Noise without Scaling
12	22222222222	$3.167\text{x}10^{-5}$	$2.609\text{x}10^{-5}$
12	332222222	$5.179\text{x}10^{-5}$	$2.320\text{x}10^{-5}$
12	43222222	$5.092\text{x}10^{-5}$	$2.211\text{x}10^{-5}$
12	54322	$5.028\text{x}10^{-5}$	$2.157\text{x}10^{-5}$
13	222222222222	$3.167\text{x}10^{-5}$	$2.609\text{x}10^{-5}$
13	333222222	$5.167\text{x}10^{-5}$	$2.319\text{x}10^{-5}$
13	443322	$5.051\text{x}10^{-5}$	$2.209\text{x}10^{-5}$
13	5443	$5.028\text{x}10^{-5}$	$2.157\text{x}10^{-5}$
Quantization Noise		$3.860\text{x}10^{-4}$	$3.860\text{x}10^{-4}$

desired, the quantization noise forces us to use configurations with minimum stage resolution.

As we decrease the desired SNR, lower resolutions may be more advantageous. In that case, the topology can be flash or pipeline. In this case, the pipeline ADC gives better results, considering the power and the area. However, pipeline architecture cannot compete with the delay and speed advantage of the flash. If high sampling rates are desired, then the pipeline architecture requires very large slew rate values. These high values can be obtained for a comparator but they may become very costly for an amplifier (generally an OTA in pipeline).

The flash architecture and the pipeline architecture are compared in Table 2.14. The number of comparators itself may be sufficient to show the high area and power ratios. On the other hand, if high speed and minimum delay is a must, the flash architecture promises better performance.

TABLE 2.14 Comparison of Architectures for 8 Bits.

Type	Configuration	No. of Amplifiers	No. of Comparators	Stages Requiring Calibration	Delay (clockcycle)
8-bit pipeline	2222222	7	21	2	7
8-bit pipeline	3332	4	23	1	4
8-bit pipeline	442	3	33	1	3
8-bit flash	-	-	225	-	1

In this design, the methodology can find a solution from the design space created by the library elements. However, this is not the case every time. If a solution could not be found by searching the library elements, an input sweep operation can be executed.

Each parameter that will be used for the sweep process introduces another dimension. Hence, some parameters are kept constant to reduce the time spent by the tool. In this example, the sampling rate was kept constant. This tool is also useful for observing the effect of the parameters on different designs.

References

[1] Van der Plas, G., et al., EsteMate: a tool for automated power and area estimation in analog top-down design and synthesis, *Proc. of IEEE 1997 CICC*, 139, 1997.

[2] Harjani, R. and Shao, J., Feasibility and performance region modeling of analog and digital circuits, *Analog Integrated Circuits and Signal Processing*, 23, 10, 1996.

[3] Veselinovic, P., et al., A flexible topology selection program as part of an analog synthesis system, *Proc. of European Design and Test Conference*, 119, 1995.

[4] Leenaerts, D.M.W., Application of interval analysis for circuit design, *IEEE Transactions on Circuits and Systems*, 803, 37, 1990.

[5] Nunez, A. and Vemuri, R., An analog performance estimator for improving the effectiveness of CMOS analog system synthesis, *Proc. of DATE'99*, 406, 1999. 1999.

[6] Casinovi, G. and Sangiovanni-Vincentelli, A., A macromodeling algorithm for analog circuits, *IEEE TCAD*, 150, 10, 1991.

[7] Vandewalle, J., De Man, H., and Rabaey, J., Time, frequency, and z-domain modified nodal analysis of switched capacitor networks, *IEEE TCAS*, 186, 28, 1981.

[8] Assael, J., Senn, P., and Tawfik, M., A switched-capacitor filter silicon compiler, *IEEE JSSC*, 166, 23, 1988.

[9] Robertini, A. and Guggenbuhl, W., Errors in SC circuits derived from linearly modeled amplifiers and switches, *IEEE TCAS - I*, 93, 39, 1992.

[10] Alpaydin, G., Erten, G., Balkir, S., and Dündar, G., Synthesis of switched capacitor filters in a multi-level optimization environment, *Proc. of Third IEEE International Workshop on Design of Mixed-Mode Integrated Circuits and Applications*, 175, 1999.

[11] Alpaydın, G., Erten, G., Balkir, S., and Dündar, G., Multi-level optimization approach to switched capacitor filter synthesis, *IEE Proceedings – Circuits, Devices, and Systems*, 243, 147, 2000.

[12] Sheu, B. J. and Choi, J., *Neural Information Processing and VLSI.* Kluwer Academic Publishers, Boston, 1995.

[13] Zornetzer, S. F., Davis, J. L., Lau, C., and T. M. (editors), *An Introduction to Neural and Electronic Networks.* Academic Press, San Diego, 1995.

[14] Lippmann, R. P., An introduction to computing with neural nets, *IEEE ASSP Magazine*, 4, 1987.

[15] Fausett, L., *Fundamentals of Neural Networks.* Prentice Hall, Englewood Cliffs, NJ, 1st ed., 1994.

[16] Hassoun, M. H., *Fundamentals of Artificial Neural Networks.* The MIT Press, Cambridge, MA, 1st ed., 1995.

[17] Leong, P. H. W. and Jabri, M. A., A low-power VLSI arrhythmia classifier, *IEEE TNN*, 1435, 6, 1995.

[18] Alpaydın, E. and Gürgen, F., Comparison of statistical and neural classifiers and their applications to optical character recognition and speech classification, in *Neural Networks Systems Techniques and Applications* (C. T. Leondes, ed.), pp. 61–88, Academic Press, NewYork, 1998.

[19] "Special issue on everyday applications of neural networks," *IEEE TNN*, vol. 8, pp. 825–975, July 1997.

[20] Manevendra, M., Parallel environments for implementing neural networks, *Neural Computing Surveys*, 48, 1, 1997.

[21] Lindsey, C. S. et. al., Real time track finding in a drift chamber with a VLSI neural network, *Nuclear Instruments and Methods in Physics Research-A*, 346, 317, 1992.

[22] Mundie, D. B. and Massengill, L. W., Weight decay and resolution effects in feedforward artificial neural networks, *IEEE TNN*, 170, 2, 1991.

[23] Schönauer, T., et. al., Digital neurohardware: Principles and perspectives," *Proc. MicroNeuro'98*, 1998.

[24] Dündar, G., Hsu, F. -C., and Rose, K., Effects of nonlinear synapses on the performance of multilayer neural networks, *Neural Computation*, 939–949, 8, 1996.

[25] Şimşek, A., Civelek, M., and Dündar, G., Study of the effects of nonidealities in multilayer neural networks with circuit level simulation, *Proc. MELECON96*, 1996.

[26] Dündar, G. and Rose, K., The effects of quantization on multilayer neural networks, *IEEE TNN*, 1446, 6, 1995.

[27] Dolenko, B. K. and Card, H., Tolerance to analog hardware of on-chip learning in backpropagation networks, *IEEE TNN*, 1045, 6, 1995.

[28] Piche, S. W., The selection of weight accuracies for madalines, *IEEE TNN*, 432, 6, 1995.

[29] Lont, J. B. and Guggenbühl, W., Analog CMOS implementation of a multilayer perceptrons with nonlinear synapses, *IEEE TNN*, 457, 3, 1992.

[30] Hollis, P. W. and Paulos, J. J., Artificial neural networks using MOS analog multipliers, *IEEE JSSC*, 849, 25, 1990.

[31] Edwards, P. J. and Murray, A. F., *Analogue Imprecision in MLP Training.* World Scientific Publishing, 1st ed., 1996.

[32] Frye, R. C., Rietman, E. A., and Wong, C., Back-propagation learning and nonidealities in analog neural network hardware, *IEEE TNN*, 110, 2, 1991.

[33] Edwards, P. J. and Murray, A. F., Fault tolerance via weight noise in analog VLSI implementations of MLPs–a case study with EPSILON, *IEEE TCAS - II*, 1255, 45, 1998.

[34] Öğrenci, A. S., Dündar, G., and Balkır, S., Fault tolerant training of neural networks in the presence of MOS transistor mismatches, *IEEE TCAS - II*, 272, 48, 2001.

[35] Bayraktaroğlu, I., Öğrenci, A. S., Dündar, G., Balkır, S., and Alpaydın, E., ANNSyS: An analog neural network synthesis system, *Neural Networks*, 325, 12, 1999.

[36] Murray, A. F. and Edwards, P. J., Enhanced MLP performance and fault tolerance resulting from synaptic weight noise during training, *IEEE TNN*, 792, 5, 1994.

[37] Chang, H., Liu, E., Neff, R., Felt, E., Malavasi, E., Charbon, E., Sangiovanni-Vincentelli, A., and Gray, P., Top-Down, Constraint-Driven Design Methodology Based Generation of n-bit Interpolative Current Source D/A Converters, *Proc. of Custom Integrated Circuits Conference*, 369, 1994.

[38] Medeiro, F., Perez-Verdu, B., and Rodriguez-Vazquez, A., Top-Down Design of High-Performance Sigma-Delta Modulators, Kluwer Academic Publishers, Dordrecht, Netherlands, 1999.

[39] Brauns, G. T., Bishop, R.J., Steer, M.B., Paulos, J.J., and Ardalan, S.H., Table-based modeling of Δ-Σ modulators using ZSIM, *IEEE Transactions on Computer-Aided Design*, 142, 9, 1990.

[40] Williams, L. A. and Wooley, B., MIDAS – a functional simulator for mixed digital and analog sampled data systems, *IEEE Intl. Symp. on Circ. and Syst.*, 2148, 1992.

[41] Liberali, V., Dias, V., Ciapponi, M., and Maloberti, F., TOSCA: A Simulator for Switch-Capacitor Noise-Sharping A/D Converters, *Proc. of the IEEE International Symposium on Circuits and Systems*, 2677, 1991.

[42] Jusuf, G., Gray P., and Sangiovanni-Vincentelli, A., CADICS–Cyclic Analog-to-Digital Converter Synthesis, *Proc. IEEE ICCAD* 286, 1990.

[43] Peralias, E., Acosta, A.J., Rueda, A., and Huertas, J.L., VHDL-based Behavioral Description of Pipeline ADC's, *Proc. of the IEEE International Symposium on Circuits and Systems*, 681, 2000.

[44] Degrauwe, M. G., Nys, O., Dijkstra, E., Rijmenants, J., Bitz, S., Goffart, B.L.A, Vittoz, E.A., Cserveny, S., Meixenberger, C., Stappen G.V., and Oguey, H.J., IDAC: An Interactive Design Tool for Analog CMOS Circuits, *IEEE Journal of Solid-State Circuits*, 1106, SC-22, 1987.

[45] Jusuf, G., Gray, P., and Sangiovanni-Vincentelli, A., A Performance-Driven Analog-to-Digital Converter Module Generator, *Proc. of the IEEE International Symposium on Circuits and Systems*, 2160, 1992.

[46] Sabiro, S., Semi, P., and Tawfik, M., HiFADiCC: A Prototype Framework of a Highly Flexible Analog to Digital Converters Silicon Compiler, *Proc. of the IEEE International Symposium on Circuits and Systems*, 1114, 1990.

[47] Vital, J. and Franca, J., Synthesis of High-Speed A/D Converter Architectures with Flexible Functional Simulation Capabilities, *Proc. of the IEEE International Symposium on Circuits and Systems*, 2156, 1992.

[48] Harjani, R., Rutenbar, R.A, and Carley, L.R., OASYS: A Framework for Analog Circuit Synthesis, *IEEE Transactions on Computer-Aided Design*, 1247, 8, 1989.

[49] Gielen, G. E. and Franca, J.E., CAD Tools for Data Converter Design: An Overview, *IEEE Transactions on Circuits and Systems II*, 77, 43, 1996.

[50] Nuno, C., Horta, N., and Franca, F., Algorithm-Driven Synthesis of Data Conversion Architectures, *IEEE Transactions on Computer-Aided Design*, 1116, 16, 1997.

[51] Lewis, S. H., Optimizing the Stage Resolution in Pipelined, Multistage, Analog-to-Digital Converters for Video-Rate Applications, *IEEE Transactions on Circuits and Systems. II*, 516, 39, 1992.

[52] Yotsuyanagi, M., Toshiyuki, E., and Kazumi, H., A 10-b 50-MHz Pipelined CMOS A/D Converter with S/H, *IEEE Journal of Solid-State Circuits*, 292, 28, 1993.

[53] Goes, J., Vital, J.C., and France, J.E., Systematic Design for Optimization of High- Speed Self-Calibrated Pipeline A/D Converters, *IEEE Transactions on Circuits and Systems. II*, 1513, 45, 1998.

[54] Soenen, E. G. and Geiger, R.L., An Architecture and an Algorithm for Fully Digital Correction of Monolithic Pipelined ADC's, *IEEE Transactions on Circuits and Systems. II*, 143, 42, 1993.

[55] McNutt, M. J., LeMarques, S., and Dunkley, L., Systematic Capacitance Matching Errors and Corrective Layout Procedures, *IEEE Journal of Solid-State Circuits*, 611, 28, 1994.

[56] McCreary, J. L., Matching Properties, and Voltage and Temperature Dependence of MOS Capacitors, *IEEE Journal of Solid-State Circuits*, 608, SC-16, 1981.

[57] Shyu, J., Temes G.C., and Krummenacher, F., Random Error Effects in Matched MOS Capacitors and Current Sources, *IEEE Journal of Solid-State Circuits*, 948, SC-19, 1984.

[58] Goes, J., Vital, C., and Franca, J.E., A CMOS 4-bit MDAC with Self-Calibrated 14-bit Linearity for High-Resolution Pipelined A/D Converters, *Proc. of IEEE Custom Integrated Circuits Conference*, 105, 1996.

[59] Kwok, P. T. F. and Luong, H.C., Power Optimization for Pipeline Analog-to-Digital Converters, *IEEE Transactions on Circuits and Systems II*, 549, 46, 1999.

[60] Sumanen, L., Waltari, M., and Halonen, K.A.I., A 10-bit 200-MS/s CMOS Parallel Pipeline A/D Converter, *IEEE Journal of Solid-State Circuits*, 1048, 36, 2001.

Chapter 3

Circuit Level Synthesis

3.1 Introduction

Modern applications such as neural information processing systems, speech and pattern recognition systems, and portable information appliances for accessing the information superhighway require analog circuitry and should leverage the power of analog design in order to remain feasible and cost-effective. Most of the knowledge-intensive and challenging design effort spent in such system designs is due to analog building blocks, whereas the digital counterparts are now synthesized rapidly in a top-down manner using widely accepted Computer-Aided Design (CAD) tools. Motivated by the potential gains arising from the reduced time to market, the development of CAD tools that automate and speed up the design process of analog circuits has become an active research area in both industry and academia. There are a number of common goals behind every existing analog circuit design automation environment. These can be listed as follows:

- *Fast and accurate development of analog building blocks*: As opposed to knowledge-intensive manual circuit design, which can become quite tedious, automation of the design can significantly reduce the time to market. In addition, accuracy of the first time design becomes an inherent feature of the automation task that cannot be guaranteed in the case of manual designs.

- *Compatibility of the end results with those of the experienced designer*: The performances of automatically designed blocks should match the ones that are obtained by careful full-custom designs.

- *Easy adaptation and interfacing*: A user-friendly interface and a reasonable design preparation overhead are key features that permit the designer to focus on system level issues without getting overwhelmed with the details of the analog circuit design process. Moreover, a desirable feature of a circuit design automation environment is its adaptability into a larger system level automation environment.

Over the years, the design automation community adopted the name *synthesis* in referring to the task of fully automatic design of an analog circuit with a predetermined qualitative behavior. A circuit-level analog synthesis system takes in the block performance specifications and outputs the transistor *sizing* and *biasing* information of a particular circuit topology that meets the nominal specifications. The focus of this chapter is on automatic circuit-level synthesis, which is the second step in the analog VLSI design automation flow of Figure 1.4.

The synthesis of high-performance analog ICs is complicated by the design objectives and performance constraints that are generally non-linear functions of circuit parameters. This inherent non-linearity renders the problem of finding a global design solution that meets the performance requirements a difficult one. Hence, an automatic synthesis strategy and the corresponding design tool that relies on this strategy should provide the following performance metrics for achieving a global solution.

- *Degree of Automation:* This is the most important metric in addressing the goals of fast and accurate development of circuits with reduced design costs and time to market. There are three significant aspects of automation: *preparatory overhead, synthesis time,* and *starting point independence.*

   Preparatory overhead is a measure of the effort and consequently, the time required for the designer to transform the circuit design requirements into certain specifications ready for input to the synthesis tool. Setting up certain performance equations and formulating the problem for optimization are fundamental components of this overhead.

   Synthesis time is the actual time consumed during the synthesis task that follows the preparatory phase.

   Starting point independence is a crucial aspect of a general analog synthesis environment. As opposed to an optimization tool, which starts from an externally supplied initial set of search variables, the synthesis tool should be able to start iterating on a given circuit design without externally specifying an initial search variable set.

- *Accuracy:* The synthesis tool should be able to predict the performance of the finalized design as close as possible to that of the fabricated one.

- *Variation-tolerance:* To be practical for the circuit designer, the synthesis tool should address the effects of manufacturing process variations on the overall circuit performance during design iterations.

- *Generality:* The tool should handle a broad range of high-performance analog circuit design problems.

- *Fabrication process independence:* Modern IC fabrication processes are frequently being subjected to shrinking. To address this issue, the synthesis strategy should accommodate the mapping of existing designs to newer technologies by incorporating the target manufacturing process parameters as externally supplied inputs.

3.1.1 Survey of Existing Synthesis Approaches

Every approach requires the task of solving a set of physical circuit equations that transform device sizes and bias values into performance parameters. However, this task is generally very complicated due to non-linear relationships between design objectives and circuit design variables. The existing analog circuit synthesis tools can be classified based on how they perform sizing and biasing of an analog circuit that meets the nominal specifications. In this respect, techniques for analog circuit design automation that appeared in the literature within the last two decades can be classified in two main categories: *knowledge-based* and *optimization-based* [1]. In the knowledge-based approach, the analog designer generates synthesis rules with the expert knowledge and incorporates them into an algorithmic procedure that leads to a design solution [2–5]. The optimization-based approach uses a search algorithm, which forces the design toward a solution specified by the user [6–7]. Further improvements have been achieved by combining optimization techniques with expert systems or model based constraints in [8–10]. These two approaches are displayed in their basic form in Figure 3.1.

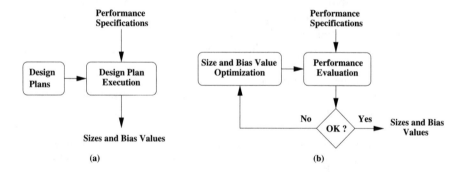

(a) (b)

FIGURE 3.1

The two distinct analog circuit synthesis approaches: (a) the knowledge-based approach, (b) the optimization-based approach.

i) Knowledge-Based Approaches: Early efforts in 1980s on analog circuit synthesis were knowledge-based. Substantial human intervention in gathering design knowledge for a particular circuit topology was required in these approaches. Once the circuit design equations and strategies were developed, they were cast into a computer program for execution in a synthesis run. The final design solution was obtained directly using this program which took the circuit specifications as inputs. We present several tools based on this approach below.

- **IDAC**: This tool was developed by researchers at the Centre Suisse d'Electronique et de Microtechnique in the early 1980s [2]. To perform sizing, IDAC employed design plans that were manually derived. These plans require inverted circuit analysis equations to solve for transistor sizes that would yield the required performance specifications. Hard-to-formalize heuristics and simplifications were employed in creating the plans. For a given circuit topology with a pre-loaded design plan, the tool produced an initial design that was fine-tuned later with gradient-based numerical optimization. A major advantage of this tool is the fast-performance space exploration capability due to compiled design plans. In addition, the tool inherently captures the expertise of analog designers during the circuit library compilation phase. It is reported that this system can handle the sythesis of circuits from 15 distinct classes such as opamps, comparators, voltage references, etc. However, this approach suffers from a demanding preparatory overhead in developing fast-executing design plans for each topology, as well as a lack of flexibility in design plan creation, which limits the spectrum of circuits available from the design library.

- **OASYS**: As opposed to IDAC, this tool decomposes a circuit to be synthesized into subblocks such as current mirrors, differential pairs, etc [3]. Although this tool also employed manually derived design plans, these plans were treated as part of a subblock within a circuit design hierarchy. Also built into the synthesis flow were a backtracking mechanism and a heuristic topology selection technique. While backtracking up the hierarchy ensured a functional design by testing different subblock configurations, alteration of the topology selection was simultaneously performed through the hierarchy until the device level. Another advantage created by hierarchy was the design reuse of subblocks in higher-level blocks. Despite the aforementioned advantages, the time-consuming design plan creation and the heuristics involved limited the application areas of this tool.

- **BLADES**: This tool is based on artificial intelligence (AI) techniques in encoding design knowledge for synthesis [5]. The approach taken in this work used both formal and intuitive knowledge in the design process within a rule-based expert system framework. A divide-and-conquer strategy was implemented in the BLADES design environment to handle a wide range of subcircuit functional blocks.

- **ISAID**: This tool also employs artificial intelligence techniques and performs sizing and biasing in two steps [11]. In the first step, the system performs a hierarchical synthesis similar to OASYS. Once a first-cut design is completed, subsequent adjustments to the design are done using the *qualitative reasoning* AI technique. This technique estimates how a performance specification will be affected for a change in a particular device size or bias. The circuit parameters are then fine-tuned via successive estimations in meeting the overall design specifications.

To summarize, knowledge-based synthesis approaches rely on time-consuming heuristic design techniques, which are also difficult to formalize. As a result, the pertaining preparatory overheads were too large. Moreover, the applicability of these approaches to general industry-grade circuit designs was also limited. Hence, these first-generation approaches could not win success in the commercial market.

ii) Optimization-Based Approaches: A second generation of methods were developed starting in the late 1980s, following the first-generation techniques. The motivation behind this research was to address the need for flexible circuit synthesis tools with reasonable preparatory overheads. Tools in this category formulate the analog circuit design problem as an optimization problem and employ numerical optimization techniques to minimize a set of objective functions while satisfying a set of specification constraints. Both objective functions and constraints are generally treated as non-linear functions of a set of independent variables — sizes of devices, values of passive components, and bias values — of the analog circuit synthesized.

The optimization-based approach is shown in Figure 3.1b. Any tool belonging to this category consists of an optimization- driven search module and an evaluation module. This is because, at every iteration of the search, the performance of the circuit needs to be evaluated for proper convergence to desired specifications. There are two commonly adopted subcategories of performance evaluation: *equation-based* and *simulation-based*, whereas optimization-driven search techniques can categorized as *classical* and *global*. We delineate these categories below.

- **Equation-Based Evaluation:** The mechanism for evaluating the performance of a circuit being optimized is based on analytically derived design equations. These performance equations are evaluated at every pass of design space exploration in order to guide the synthesis flow toward the design specifications. Two major advantages of employing equations in the evaluation phase stem from inherently fast evaluation times and added flexibility in choosing the search variables. In addition, a breakthrough in automatic derivation of design equations came with the advent of symbolic simulation techniques. These techniques automatically generate small-signal transfer functions of arbitrary circuit topologies for AC behavior characterization. While equation-based approaches allow easier incorporation of expert information, there are major drawbacks to this approach; the pertaining design equations still need to be derived accurately and symbolic analysis techniques suffer from exponential growth in complexity of function generation with the circuit size. These drawbacks render the equation-based approaches inefficient and inaccurate for large circuits targeted for submicron technologies.

- **Simulation-Based Evaluation:** To alleviate the problems that plague equation-based approaches, simulation-based evaluation became popular over recent years due to the availability of ubiquitous and fast computing power. Synthesis tools belonging to this category perform some form of numerical circuit simulation to evaluate the performance at every iteration of the search cycle. Depending on how extensive the simulation methodology is, the approach may significantly reduce the design equation derivation overhead. Major advantages of simulation-based approaches can be outlined as high accuracy in evaluation, generality in terms of handling a broad range of circuits, and low preparatory effort. However, these approaches still need to improve on certain aspects such as their dependence on the availability of highly accurate circuit simulation models for different technologies, and the need to perform multiple simulations to fully characterize a circuit in different domains (i.e., AC and DC). Moreover, the state-of-the-art semiconductor fabrication technologies require the designs to be tolerant to manufacturing variations, which is yet another demanding issue that needs to be addressed.

- **Classical Search:** Algorithms that typify classic search are based on steepest descent following gradients and quadratic programming techniques. A major disadvantage of classical search techniques is their fast convergence to frequently encountered local minima in search spaces. Although a good starting point may alleviate this problem, creation of starting points cannot be always sustained by the users of synthesis tools, especially for complex circuit topologies.

- **Global Search:** Aggressive search techniques with global convergence have became more popular among the developers of synthesis tools. Typical techniques for global search are branch-and-bound algorithms, simulated annealing, and genetic algorithms. Branch-and-bound algorithms can provide efficient solution space exploration, however, they may get extremely slow due to exponential growth with circuit size. Simulated Annealing (SA) on the other hand, is a popular technique due to its global convergence ability. SA is an effective technique in simultaneously handling both continuous and discrete search spaces, but it can fail in converging to global minimum unless extreme care is exercised in choosing an appropriate cooling schedule and a suitable termination criterion. Genetic algorithms (GA) are also suitable candidates for global convergence. Disadvantages of GAs can be summarized as slow convergence and heuristics involved in choosing a particular genetic methodology, selection criteria, population size, chromosome encoding, etc.

While a majority of synthesis tools can be placed in a definite search/evaluation category combination, recent tools use hybrid techniques in an effort to combine the advantages of various approaches. Table 3.1 below displays a matrix of widely known optimization-based synthesis tools from the literature with respect to their search and evaluation techniques.

TABLE 3.1 Optimization-Based Synthesis Environments.

Evaluation	Classical Search	Global Search
Simulation	DELIGHT.SPICE	FRIDGE ANACONDA/MAELSTROM OAC
Equation	OPASYN STAIC Maulik	AMGIE GPCAD SD-OPT VASE
Mixed		ASTRX/OBLX Ours

- **DELIGHT.SPICE:** Developed at the University of California-Berkeley, this is an interactive system designed to fine-tune an existing circuit toward performance specifications [7]. The tool employs a gradient-based technique named *method of feasible directions*, where a subset formed of worst performance and constraint functions direct the search. For the evaluation part, the tool uses the SPICE circuit simulator. The inefficiency of this approach is due to the good initial starting point requirement and inherently slow optimization speed caused by employing SPICE in the evaluation cycle.

- **FRIDGE:** This tool also uses SPICE for the evaluation cycle [12]. To address the initial condition requirement, it employs a simulated annealing-based global optimization technique. Although the number of SPICE iterations are kept at a minimum by a fast cooling schedule, this tool still suffers from inherently long execution times.

- **ANACONDA/MAELSTROM:** These tools use a global optimization algorithm based on stochastic pattern search with inherent parallelism [13], [14]. This parallelism has been exploited in these tools by distributing the computational loads of search and evaluation across a network of parallel workstations. This tool has been successfully applied to industry-grade designs, but the run times are still regarded long in general.

- **OAC:** This is a non-linear optimization tool tailored specifically to operational amplifier synthesis [15]. The tool exploits previously designed circuits stored in its database and redesigns an existing circuit for a new set of specifications.

- **OPASYN:** This is one of the earlier tools that combine optimization-based approaches with equation-based evaluation [6]. While the design equations were manually derived, the search space was explored by steepest descent algorithm. To address the initial point problem, multiple starting points were selected. The tool can run substantially faster than simulation-based counterparts, however, it requires significant human intervention in deriving the circuit design equations.

- **STAIC:** Similar to OPASYN, STAIC also uses an equation basis for evaluating the performance [16]. The novelty of the tool is a two stage optimization algorithm. In the first stage, simple circuit equations are used to explore the design space. Next, a refinement of the solution is performed using more accurate models. Design expertise is still a fundamental component of this environment.

- **Maulik:** This is a relatively recent equation-driven synthesis environment, where industry standard complex device simulation models are also handled without resorting to simulation [17]. This has been made possible in this tool by adopting a *relaxed DC formulation* of the circuit under consideration. As opposed to verifying the DC operation of the circuit in each evaluation cycle, the approach taken in this work is to include DC solutions as part of the constraint functions. Hence, the accuracy of the DC points is regarded as a constraint. The system uses quadratic programming techniques for the search cycle. Substantially complex circuits have been successfully synthesized with this tool, although good initial starting point and circuit design equation derivation remain as disadvantages.

- **AMGIE:** Developed at Katholieke Universiteit at Leuven, Belgium, this is a complete second generation synthesis system with an equation-based evaluation cycle with global search [18]. A key feature of this environment is the automatic derivation of small-signal transfer functions by symbolic analysis. Any type of circuit topology can be handled by this approach with the inclusion of user-supplied large-signal circuit equations. However, a major disadvantage of this approach is the exponential growth in transfer function with circuit complexity. Although the system employed pruning techniques for simplification, this results in the degradation of evaluation accuracy. The fact that AMGIE uses simulated annealing as its optimization algorithm also compounds the problem of slow execution times.

- **GPCAD:** This is a specialized tool for the design of CMOS opamps, which formulates the design problem as a posynomial convex optimization problem [19–20]. The employed optimization technique is geometric programming that can generate designs very fast. However, similar to other equation-based evaluation approaches, this tool also requires the derivation of design equations.

- **SD-OPT:** This tool has been developed for the automated design of $\Sigma\Delta$ modulators [21]. The simulated annealing-based optimizer generates the subblock specifications of the converter, whereas the evaluation cycle uses symbolic equations describing the converter architecture under consideration.

- **VASE:** This tool presents one of the early attempts on synthesis of analog systems from a VHDL-AMS (analog-mixed signal) behavioral description [22]. The environment is based on a two-level optimization approach that ultimately produces sized subblocks. Alternative topologies are first generated by a branch-and-bound algorithm. Next, a genetic algorithm based technique is employed to size the subblocks and fixes the topologies. For fast execution, analytic design equations are utilized during the performance evaluation cycle.

- **ASTRX/OBLX:** Developed at Carnegie Mellon University, this is a mixed simulation-equation approach [9]. Instead of SPICE, the asymptotic waveform evaluation (AWE) technique has been devised for small-signal representations. While AWE has substantial speed-up over conventional simulators, all other types of evaluation still need to be provided by the user. Simulated annealing has also been the choice of optimization in this system combined with a relaxed DC formulation for improved efficiency. ASTRX/OBLX has shown success in synthesizing complex designs although execution times remained large.

In view of this literature survey on state-of-the-art in analog synthesis, it can be conjectured that simulation dominant approaches with SPICE level accuracy are required for handling industry-grade designs. Moreover, for applicability to a broad range of circuits, the overhead of representing certain circuit parameters with equations should be kept at a minimum.

Finally, it should be noted that there is renewed interest from the IC manufacturing industry for yield maximization and manufacturability of designs. While the majority of the analog synthesis environments focus only upon optimizing nominal designs, an industry-standard analog design automation system should also address issues related to robustness and yield [23–24]. Thus, the incorporation of tolerance to statistical process variations and mismatches has recently become an integral part of these systems. An earlier attempt to incorporate manufacturability has been presented in [25]; ASTRX/OBLX tool has been augmented with a search mechanism for the worst case corners. In [26], closed-form expressions for the sensitivities of the performances to the process parameters have been employed in the optimization iterations, which allowed the simultaneous synthesis of circuits for both manufacturability and nominal performance. A unified method based on the generalized boundary curve for nominal design and design centering for yield maximization is presented in [27], where the circuit optimization problem is based on two robustness objectives defined as parameter distances for the nominal design and worst-case distances for the design centering [28]. A recent analog circuit synthesis tool incorporating mismatch is presented in [29]. This is a simulation-based tool that includes tolerance analysis, performance optimization, and design centering.

3.1.2   Evolutionary Computation for Analog Synthesis

Research efforts driven by evolutionary computation techniques have begun to appear in the literature over the last few years. The use of genetic algorithms (GAs) to select filter component sizes and filter topologies are presented in [30–31]. A comparison of genetic-based techniques applied to filter design is presented in [32]. In the work of Koza et al. on analog circuit synthesis by means of genetic programming (GP), the component values, number of components, and the circuit topologies are evolved together, where various analog filter design and circuit synthesis problems have been solved using this approach [33]. The capability to evolve circuits of moderate complexity using a GA based on a simple linear circuit representation is demonstrated in [34]. Research efforts for the exploitation of evolutionary techniques also involve Field Programmable Analog and Transistor Arrays. In this category of analog synthesis, researchers mostly focus on automatic reconfiguration of field programmable devices driven by Evolutionary Computation techniques [35–36]. Although there has been considerable interest in the application of evolutionary techniques to analog synthesis, most of the reported work in the past concentrates on filter applications and evolving circuits composed of discrete components.

Due to their ability to find a satisfactory solution in a short time, GAs have been recently employed as optimization routines for analog circuit design in both industry (for example: *Neolinear, Inc., Analog Design Automation, Inc.*) and academia [37]. A combined genetic/annealing algorithm for simulation-based synthesis of custom analog cells has been presented in [14]. The approach taken there was based on distributing the load of a serial annealer to parallel computational nodes for speed-up. An automated circuit design system for the evolution of CMOS amplifiers was recently presented in [38]. The proposed technique is based on a combination of genetic programming and a topology independent optimization method using current-flow analysis. Hence, the applicability of evolutionary computation techniques within the framework of an analog synthesis CAD tool targeting mainstream VLSI technologies and analog integrated circuit (IC) design remains an active research area. Moreover, such a tool should be able to handle a broad range of analog design problems and synthesize variation-tolerant high-performance circuits.

In this chapter, a new analog IC synthesis tool driven by a novel evolutionary optimization technique is presented. The synthesis strategy is capable of fully automating the design flow, starting from a circuit topology and performance specifications and ending at a circuit with appropriate transistor sizes, bias values, and passive component magnitudes. Moreover, a mechanism for establishing tolerance to manufacturing process variations has also been implemented in the synthesis flow to attain robust operation of fabricated circuits. Hence, in an effort to address the general shortcomings of mainstream techniques the presented analog synthesis tool has the following key features:

- An unconstrained optimization formulation to which the circuit synthesis problem is mapped.

- A novel high performance search algorithm based on combining Evolution Strategies (ES) and Simulated Annealing (SA), yielding higher computational efficiency and improved optimization speed when a large set of various circuit topologies is considered.

- A fast DC circuit simulator that works in the inner loop of the optimization algorithm for the calculation of DC operating point parameters that are then used in the DC constraint generation and for the verification of the DC operating point.

- User-defined equations complemented with neural performance models that enhance the flexibility of the optimization of AC parameters for arbitrary circuit topologies without resorting to computationally intensive AC simulations.

- Incorporation of the real-life IC manufacturing technology parameter variations and the resulting mismatch effects into the synthesis loop through neural-fuzzy models for robust operation after fabrication.

The synthesis system has been completely implemented in software as a number of interconnected tools and interfaced between the system and layout level modules of the design automation flow depicted in Figure 1.4. To show the viability of the evolutionary techniques within the framework of a complete analog CAD tool, various high-performance ICs with tight design specifications and constraints have been synthesized successfully for manufacturing variation tolerance. Moreover, a prototype test chip including the automatically synthesized analog circuits has been fabricated in CMOS technology and the test results have demonstrated that the design specifications have also been achieved on silicon.

In the next section, the automatic synthesis strategy for analog ICs is outlined followed by the novel evolutionary optimization technique employed in the automatic synthesis environment. Various analog IC synthesis examples and their results are given at the end of the chapter.

3.2   Evolution-Based Automatic Synthesis Strategy

The structural diagram of the proposed synthesis architecture that maximizes the performance metrics discussed in Section 3.1 is shown in Figure 3.2.

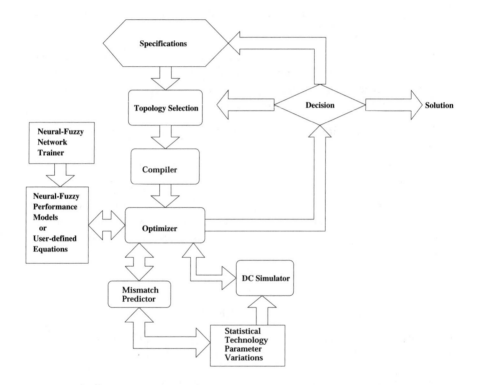

FIGURE 3.2
The synthesis architecture.

The presented architecture is part of a larger design automation environment and topology selection is done outside of this architecture. Hence, starting from a given topology, the synthesis architecture is composed of the following key components.

- **Compiler:** For a given topology, the first task in the synthesis process is to define the search variables, specifications, and constraints in an appropriate manner. The compiler module realizes this task by translating the design specifications and constraints defined by the user into a cost function whose minimum corresponds to the circuit design parameters that best match the given specifications. The structure of the cost function is described in the next section.

- **Optimizer:** Circuit optimization refers to choosing circuit parameters that result in the best circuit performance with respect to design specifications and constraints. These specifications and constraints may exist in the form of an input netlist defined by the user. Considering a CMOS analog IC cell, the dimensions of the circuit components (MOS transistor channel widths W, lengths L, sizes of passive components such as capacitors, etc.), DC bias voltages, and currents are treated as search variables of the optimization problem. The mechanism of the evolution-based optimization technique employed is presented in the next section.

- **DC Simulator:** To increase the accuracy of the synthesis flow and guarantee the true optimum solution while satisfying the synthesis objectives, a fast DC circuit simulator has been used [39]. This is a direct method simulator that essentially solves a set of non-linear algebraic equations formed by combining non-linear device circuit models (i.e., MOS transistors, diodes etc.) and circuit topology equations (Kirchoff's Laws). The resulting equations are based on static DC operating conditions, which do not take into account any transient or dynamic energy storage effects of components such as capacitors and inductors. In particular, typical electronic circuit parameters acting as DC constraints such as offset voltage; DC power consumption; and objective functions such as a DC open-loop gain equation, and the bandwidth; are obtained by this simulator. Moreover, the dominance of technology parameter variations on the DC operating point and overall circuit performance requires the simulator to incorporate reliable device models for accurate representation. The simulator utilized in this work is SPICE-like but faster, since it uses previous solutions as the initial values for fast convergence [40–41].

- **Performance modeling:** Evaluating the performance of the circuit being optimized is the most crucial step of an analog IC optimizer. DC simulation provides the necessary circuit parameters for calculating the DC constraints that become part of the cost function being minimized. On the other hand, certain AC parameters such as bandwidth and gain-bandwidth product are also required in most of the synthesis applications as part of the cost function. While AC circuit simulation can provide these parameters by performing a sinusoidal steady-state analysis on the circuit under consideration at different frequencies, it also increases the computational complexity of the synthesis environment. To circumvent this problem, both user-defined performance equations and neural-fuzzy performance models are utilized instead for flexibility and speed in our approach.

- **Mismatch Prediction:** The manufacturing process variations should be incorporated into the analog IC synthesis flow for achieving variation-tolerance. The effects of these variations emerge in the form of electrical parameter mismatches between two identical transistors located at different places on the die or wafer. This may lead to considerable performance degradation in most of the modern IC applications, since matched transistor pairs are pervasive in many analog building blocks. Although there has been extensive work in the literature for mismatch characterization, incorporation of transistor mismatches into a complete IC synthesis flow has not been reported yet. The presented IC synthesis flow provides simultaneous mismatch minimization while the circuit parameters are being optimized. The algorithmic details of this novel technique are described in Section 3.6.

- **Statistical technology parameter variations:** These variations are externally supplied, allowing the incorporation of their effects for various fabrication processes.

3.3 Evolution-Based Analog IC Optimization

Analog IC optimization is a demanding process due to the existence of a large number of both discrete and continuous search variables that grows with circuit size. The non-linear relationship between these variables and the intricacy of finding performance equations complicate the optimization problem further. These disadvantages, coupled with the different nature of each analog IC cell topology, require an efficient IC optimization algorithm to have the following properties:

- The optimization algorithm should be independent of the initial variables; randomly chosen initial variables that satisfy a user-defined constraint should be adequate for convergence.

- The optimization algorithm should be independent of the circuit topology under consideration.

- The optimization algorithm should handle any kind of cost function; discontinuities should be accommodated.

- The optimization algorithm should overcome local minima while satisfying the objectives and constraints.

- The evaluation cycle should include DC constraints to avoid convergence to an unrealistic solution, which requires the evaluation of the DC constraints to be highly accurate.

The ES algorithm is a global search mechanism with a stochastic nature in mutation and random recombination ability, and exhibits the self-adaptation of parameters. Moreover, ES is a universal search method, since it does not use the derivative of the objective function. The major drawback of this algorithm is that it may reach local minima very fast and may get stuck there in case of an inappropriate selection mechanism. The selection method employed in ES must be carefully chosen to avoid convergence to local minima [42]. For instance, a uniform random selection algorithm distributes the selection randomly over the whole population and hence spends a lot of time in sectors of the solution space that are of no use. On the other hand, a selection mechanism that concentrates on the best region in the search space may find the local solution quickly and never find the global one. To mitigate these effects, fitness functions can be utilized for defining selection probability. However, the facts that these functions are problem-specific and that the performance of search strictly relies on how well qualified they are preclude the automation task.

The presented optimization approach uses an ES/SA combination in finding a satisfactory solution while satisfying the properties shown above. Moreover, a way is found for specifying a fitness function without the need for detailed expert knowledge or trial-and-error testing. This is achieved by borrowing the stochastic Metropolis criterion from SA and using it as the selection mechanism of ES. Various IC optimization experiments shown in Section 3.7 have demonstrated that both discrete and continuous variables can be updated in this approach, making it amenable to large analog IC optimization problems.

The problem-specific components of the algorithm are as follows:

- *Circuit representation*, which determines how the present state of the topology being optimized is mapped as an evolving population.

- *Cost function*, which determines how the cost of each topology is calculated.

- *Genetic operators and mechanisms* used in the algorithm to generate candidate solutions (*reproduction, recombination, mutation, selection*).

3.3.1 Circuit Representation

Assuming the topology for a CMOS analog IC cell is chosen, the initial task of the optimization tool is to define search variables, specifications, and constraints in an appropriate manner. Since ES has been used in the proposed algorithm as a dynamic step length controller, the search variables and related strategy parameters constitute the input population of the algorithm. In particular, a chromosome **x** representing the search variables of the optimization problem can be defined as:

$$\mathbf{x} = [\overbrace{p_1, p_2, \ldots, p_t}^{W/L}, \overbrace{L_1, L_2, \ldots, L_t}^{L}, \overbrace{V_1, V_2, \ldots, V_k}^{V_{bias}}, \overbrace{I_1, I_2, \ldots, I_\Psi}^{I_{bias}}, \overbrace{C_1, C_2, \ldots, C_\nu}^{Capacitors}]$$

$$(3.1)$$

where, for the analog cell under optimization,

$$p_i, \quad (i = 1, \ldots, t) : W/L \text{ ratios of MOS transistors,}$$
$$L_i, \quad (i = 1, \ldots, t) : \text{channel lengths of MOS transistors,}$$
$$t : \text{number of transistors,}$$
$$k : \text{number of bias voltages,}$$
$$\Psi : \text{number of bias currents,}$$
$$\nu : \text{number of capacitors.}$$

3.3.2   Cost Function

The presented optimization strategy relies on minimizing a cost function $C(\mathbf{x})$, given as

$$C(\mathbf{x}) = \sum_{i=1}^{l} w_i \hat{f}_i(\mathbf{x}) + \sum_{j=1}^{k} w_{l+j} \, \hat{g}_j(\mathbf{x}) \qquad (3.2)$$

where $\hat{f}()$ is a set of normalized objective functions derived from performance specifications (DC gain – A_o, gain bandwidth product – GBW, phase margin – PM, slew-rate – SR, power dissipation, offset voltages, etc.) of the analog cell that the designer wishes to optimize and $\hat{g}()$ is a set of normalized user-defined penalty terms acting as constraint functions. The indices l and k in Equation 3.2 are the number of objective and constraint functions, respectively. Once the user specifies an upper boundary vu_i and a lower boundary vl_i for the ith constraint, these are used to normalize the specification's range as $\hat{f}_i(\mathbf{x}) = \frac{f_i(\mathbf{x}) - vl_i}{vu_i - vl_i}$. Similary, $\hat{g}()$ is obtained using this method. This normalization step creates a uniform range of various specifications so that the designer can now set the relative importance of competing specifications by adjusting the scalar weights w_i. Initially, greater weight values are assigned to important objectives and constraints. These weights are then adapted automatically in the annealer in order to arrive at a satisfactory solution. This is realized in the proposed technique by monitoring the value of each objective/constraint function and increasing/decreasing the weight of each function if the particular function becomes significantly higher/lower than the total contributions of objective and constraint functions. This cost function is computed for every generation obtained from the population in the optimization loop.

3.3.3 Description of the Algorithm

An adapted form of (μ,λ) evolution strategy has been used as the core of the proposed optimization algorithm. In this scheme, a candidate solution can be represented as an individual $\mathbf{a}$, which consists of search variables $\mathbf{x}$, strategy parameter set $\mathbf{s}$, and cost function $C(\mathbf{x})$. Then, the population of individuals formed by μ parents and λ descendants for generation g can be represented as $\mathbf{P}_\mu^{(g)}$ and $\mathbf{P}_\lambda^{(g)}$, respectively

$$\mathbf{P}_\mu^{(g)} = \{\mathbf{a}_m^{(g)}\} = [\mathbf{a}_1^{(g)}, \mathbf{a}_2^{(g)}, \ldots, \mathbf{a}_\mu^{(g)}] \ni \mathbf{a}_m^{(g)} = [\mathbf{x}_m^{(g)}, \mathbf{s}_m^{(g)}, \mathbf{C}(\mathbf{x}_m^{(g)})] \quad m = 1, \ldots, \mu$$
$$\mathbf{P}_\lambda^{(g)} = \{\mathbf{a}_l^{(g)}\} = [\mathbf{a}_1^{(g)}, \mathbf{a}_2^{(g)}, \ldots, \mathbf{a}_\lambda^{(g)}] \ni \mathbf{a}_l^{(g)} = [\mathbf{x}_l^{(g)}, \mathbf{s}_l^{(g)}, \mathbf{C}(\mathbf{x}_l^{(g)})] \quad l = 1, \ldots, \lambda.$$
$$(3.3)$$

The recombination and mutation processes of the ES algorithm have been used to adapt the step length of the search variables $\mathbf{x}$. In the self-adaptation of strategy parameters, the standard deviation for mutation becomes part of the individual and evolves by mutation and recombination just as the object variables do [43]. The strategy parameters control the recombination and mutation operators for self adaptation. The following pseudocode outlines the (μ,λ)-ES algorithm using the notation introduced above [44].

> **Procedure** (μ,λ)-ES;
> **Begin**
> $g = 0$;
>   initialize $\left(\mathbf{P}_\mu^{(0)} = \left\{\mathbf{x}_m^{(0)}, \mathbf{s}_m^{(0)}, \mathbf{C}(\mathbf{x}_m^{(0)})\right\}\right)$;
>   **Repeat**
>   **For** $l = 1$ **to** λ **Do Begin**
> $\mathbf{F}_l$ = reproduction $\left(\mathbf{P}_\mu^{(g)}, \rho\right)$;
> $\mathbf{s}_l$ = s_recombination $(\mathbf{F}_l, \rho)$;
> $\tilde{\mathbf{s}}_l$ = s_mutation $(\mathbf{s}_l)$;
> $\mathbf{x}_l$ = x_recombination $(\mathbf{F}_l, \rho)$;
> $\tilde{\mathbf{x}}_l$ = x_mutation $(\mathbf{x}_l, \tilde{\mathbf{s}}_l)$;
> $\tilde{\mathbf{C}}_l = C(\tilde{\mathbf{x}}_l)$
>   **End**;
> $\tilde{\mathbf{P}}_\lambda^{(g)} = \left\{\tilde{\mathbf{x}}_l, \tilde{\mathbf{s}}_l, \tilde{\mathbf{C}}_l\right\}$;
> $\mathbf{P}_\mu^{(g+1)}$ = selection $\left(\tilde{\mathbf{P}}_\lambda^{(g)}\right)$;
> $g = g + 1$;
>   **Until** stop_criterion
> **End**

3.3.3.1 Reproduction Operator

The $\rho = 2$ case has been used for reproduction. Thus, there are two parents which take part in the procreation of a single individual. The reproduction operator randomly selects these parents and forms the parent family as a 2-tuple **F**. Empirically chosen values $\mu = 12, \lambda = 2$ for the number of parents and descendants during ES iterations have yielded satisfactory performance in the optimization runs.

3.3.3.2   Recombination Operator

Intermediate recombination has been utilized by means of vectorial formation of the center of gravity from two randomly selected parents. The descendants generated in this manner include the strategy parameters in addition to search variables.

3.3.3.3   Mutation Operator

The set of strategy parameters **s** in the (μ, λ) strategy represent an n-dimensional normal distribution for mutating the individual. In particular, these parameters are formed of the vector components $\boldsymbol{\sigma}$ and $\boldsymbol{\alpha}$ that determine the variances and covariances of the n-dimensional normal distribution used for exploring the search space, respectively. The operator is applied to the recombined individuals $\mathbf{a}_l$ by first mutating the strategy parameters $\boldsymbol{\sigma}_l$ and $\boldsymbol{\alpha}_l$. Next, $\mathbf{x}_l$ is modified according to the new set of strategy parameters obtained by mutating $\boldsymbol{\sigma}_l$ and $\boldsymbol{\alpha}_l$. The sequence of mutations can be outlined as:

- **Mutation of $\boldsymbol{\sigma}_l$:**

$$\tilde{\boldsymbol{\sigma}}_l = (\sigma_1 e^{\tau \mathbf{N}(0,1)}, \ldots, \sigma_\lambda e^{\tau \mathbf{N}(0,1)}) \qquad (3.4)$$

where the log-normal operator $e^{\tau \mathbf{N}(0,1)}$ has been used for mutation [44]. During optimization runs, satisfactory performance has been empirically achieved by setting the learning parameter $\tau = 0.3$ in generating random numbers by an exponential transformation of normal distribution $\mathbf{N}(0,1)$. In addition, the value range for σ set between 1 and 6 has provided good results.

- **Mutation of $\boldsymbol{\alpha}_l$:**

$$\tilde{\boldsymbol{\alpha}}_l = (\alpha_1 + \mathbf{N}(0, \beta^2), \ldots, \alpha_\lambda + \mathbf{N}(0, \beta^2)) \qquad (3.5)$$

The value used for β determines the rotation angles of the mutation hyperellipsoid that describes the mutations governed by n-dimensional correlated normal distribution. Empirically, $\beta = 0.0873$ corresponding to a 5° rotation has shown to yield good results whereas the value range for α has been kept between 0 and 1 [43].

- **Mutation of search variables:**

$$\tilde{x}_l = (x_1 + cor_1(\tilde{\sigma}_l, \tilde{\alpha}_l), \ldots, x_\lambda + cor_\lambda(\tilde{\sigma}_l, \tilde{\alpha}_l)) \tag{3.6}$$

where **cor** is a random vector with normally distributed and eventually correlated components. This vector is calculated by multiplying a random vector by the covariance matrix defined by $\tilde{\alpha}_l$.

3.3.3.4 Selection and Stop Criterion

The incorporation of SA as a selection mechanism can be realized by arranging the annealing temperature T as a controller, which is reduced for every ES iteration. If $\mathbf{C}_l$ and $\tilde{\mathbf{C}}_l$ are symbolized as the cost functions of search variables at the previous and current ES iterations, respectively, then the Metropolis Criterion in Equation 3.7 can be used as the selection mechanism if the new solution is worse than the old one [45]

$$random() < e^{[\frac{\mathbf{C}_l - \tilde{\mathbf{C}}_l}{T}]} . \tag{3.7}$$

Here, T is the current annealing temperature and $random()$ returns a uniformly distributed random number between $[0, 1]$. In this case, the selection mechanism of ES takes the form

$$\mathbf{select}(\tilde{\mathbf{x}}_l, \tilde{\mathbf{s}}_l) \quad \text{if} \quad T \ln(random()) + \tilde{\mathbf{C}}_l < \mathbf{C}_l . \tag{3.8}$$

In addition, if the new solution is better, it is accepted with probability 1. The creation of generations at the same temperature stops when the dynamic range of the current population falls below a tolerance value. The main evolutionary loop will then be run again at the newly updated temperature, which is controlled by a cooling schedule. The evolved population at the previous temperature is used as the parent population of the ES procedure run at the current temperature. The whole procedure terminates when $T < 0$.

3.4 DC Simulator

To increase the accuracy while maintaining a reasonable overhead, the described synthesis environment is based on a fast DC simulator. The simulator employs direct methods by solving a set of non-linear algebraic equations arising from non-linear device models (i.e. MOS transistors, diodes etc.) and topology equations of circuits subject to static DC operating conditions. This simulator has been used to generate DC constraints such as offset voltage, DC power consumption, DC penalty terms, and objective functions such as DC open-loop gain and bandwidth.

A majority of previous approaches except DELIGHT.SPICE, FRIDGE, and ANACONDA/MAELSTROM are inspired by the fact that decreasing the time of optimization can be generally achieved by eliminating the DC and AC simulators and finding near optimum solutions with models by keeping a good DC behavior with DC equation-dependent constraints. While this approach seems to be realistic for AC simulators due to computationally intensive repetitive execution requirements, the following exceptions for a DC simulator make it feasible to include it in the optimization loop.

- Compared to AC counterparts, DC simulators do not spend much time.

- Near solutions for DC operating points may cause divergence from a DC working point and consequently, may result in a malfunctioning circuit at the end of optimization.

- Technology parameter variations that affect the performance strongly are dominant on DC operating point and need to be taken into account in an accurate manner, which is possible with an accurate DC simulator.

The simulator used in the environment is SPICE-like in terms of level of accuracy but generally faster since it uses previous solutions as initial values in the optimization loop to converge more quickly. This type of fast DC convergence mechanism has been observed to be reasonable for various circuit synthesis applications because the current DC operating conditions are seldomly far away from the previous ones during optimization iterations. The simulator is currently capable of interpreting models of different complexity ranging from the basic SPICE (level=2) model to BSIM3 (Hspice level=49).

3.5 Performance Modeling

As mentioned earlier, our synthesis approach provides a flexibility in modeling the performance of circuits being optimized. This is facilitated by utilizing user-defined performance equations, neural-fuzzy performance models, or a combination of both. Here, the aim is to make use of the most efficient modeling technique for a given circuit optimization problem. For instance, a simple and well-documented analog building block such as a Basic Two Stage (BTS) opamp can be modeled via user-defined equations with a reasonable expert knowledge incorporation overhead with the DC simulator yielding the required operating point variables in the equations. However, for more complex and less documented circuits, user-defined equations may not be completely derived in a reasonable amount of preparatory overhead time. We address this case by employing neural-fuzzy performance models in this book. These two approaches are presented next.

- *User-defined performance equations:* For the case of equation-based circuit performance evaluation, the DC simulator yields the required operating point variables (device transconductances g_m and g_{mb}, output conductances g_o, etc.) and voltages used in the computation of performance equations. Once these variables are available, they are used to compute the required performance specification expressed by the pertinent user-defined equation. This mechanism is illustrated below for two examples commonly encountered in literature.

A Single-Ended Differential Amplifier

The schematic of a single ended differential amplifier is provided in Figure 3.3. Once the DC simulator generates the operating point information, power dissipation and output offset voltage of the circuit can be directly obtained. The following performance equations deemed relevant for this circuit can then be evaluated.

Specification	Equation
Output Impedance (R_{out})	$(1/g_{o4}) \| (1/g_{o2})$
DC $-$ gain (A_o)	$g_{m1,2} R_{out}$
GBW (ω_t)	$g_{m1,2}/C_L$
Output Swing High	$V ds_{M4} - V dsat_{M4}$
Output Swing Low	$V ds_{M2} - V dsat_{M2}$

A BTS Opamp

The schematic of a BTS Opamp is provided in Figure 3.4. Once the DC simulator generates the operating point information, power dissipation and output offset voltage of the circuit can be directly obtained. In addition, the following performance equations deemed relevant for this circuit can then be evaluated.

Specification	Equation
Output Impedance (R_{out})	$(1/g_{o7}) \| (1/g_{o6})$

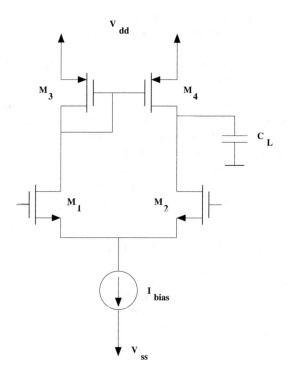

FIGURE 3.3
A single-ended differential amplifier.

DC − gain (A_o)	$g_{m1,2}\,R_{out}$
GBW (ω_t)	$g_{m1,2}/C_f$
Slew − rate (SR)	$2I_{D1}/C_f$
Output Swing High	$Vds_{M7} - Vdsat_{M7}$
Output Swing Low	$Vds_{M6} - Vdsat_{M6}$

While user-defined equations are powerful tools for performance modeling regarding circuits of moderate size, they are not able to remain feasible for large and complex circuits, mainly due to the high cost of generating complete equations and expert knowledge required for accuracy.

Neural-fuzzy performance models: An efficient alternative to reduce the preparatory overhead for arbitrarily complex circuits is to build neural-fuzzy models of the circuit's performance equations. The advantages of such a modeling effort are due to the abilities of these models to closely approximate any non-linear circuit behavior, and speed of training once the training data is generated. A three-layer neural-fuzzy network has been

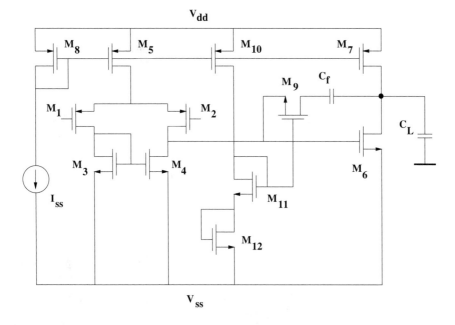

FIGURE 3.4
A BTS opamp.

employed in this work to model the analog IC's performance parameters [46]. The fuzzy logic network is shown in Figure 3.5.

In this scheme, a Gaussian shaped membership function (3.9) is used for the first layer.

$$\mu_{A_{ij}} = exp\left[-\frac{(x_i - c_{ij})^2}{2\sigma_{ij}^2}\right] \tag{3.9}$$

The second layer uses the product operation rule for fuzzy implication. Combining with Equation 3.9 the following equation is derived for the second layer that expresses rule firing strengths

$$\phi_j = \prod_{i=1}^{n} \mu_{A_{ij}}(x_i) \tag{3.10}$$

The third layer is the output layer for defuzzification process, which uses the centroid method. In this layer, the rule firing strengths and the weights corresponding to these rules are used to compute the center of gravity by

$$y = \frac{\sum_{j=1}^{m} \phi_j w_j}{\sum_{j=1}^{m} \phi_j} \tag{3.11}$$

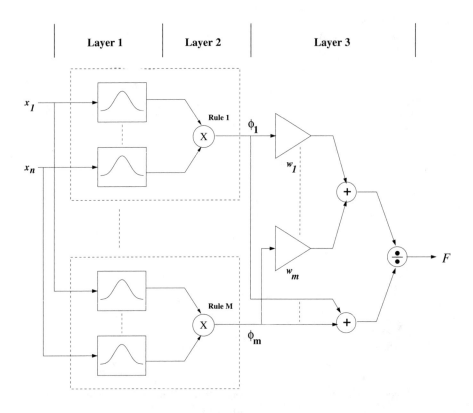

FIGURE 3.5
Three-layer adaptive neural fuzzy network.

where ϕ_j and w_j denote the position and weight of the jth output fuzzy set, respectively. This layer, also referred to as the defuzzifier, ultimately yields the decision $F(\mathbf{x}) = y$.

Considering analog circuit synthesis, the optimization variables $\mathbf{x}$ are mapped as the input variable vector of the fuzzy network. The output function F is then the performance equation we wish to model. During training for a particular equation, $\mathbf{x}$ is fed to each fuzzy rule and a value is found for F in every *epoch*. The previously described optimization algorithm is used to train the fuzzy network that minimizes the cost function $C(\mathbf{x})$, defined by:

$$C(\mathbf{x}) = w_e * e(\mathbf{x}) \qquad (3.12)$$

Here, the normalized error function is given as

$$e = \sum_{i=1}^{n} \sum_{j=1}^{m} \left(\frac{y_{ij} - y_{ij}^d}{2\lambda_j(y_{jmax} - y_{jmin})} \right)^2 \qquad (3.13)$$

where y^d is the desired output and y is the current output of the fuzzy network (considering the single-output case and supervised learning). Both y^d and y are functions of the topology of the fuzzy network. m is the number of outputs, n is the number of the input patterns used for learning. λ_j is the jth output learning rate. The key to the formulation is that the minimum of $C(\mathbf{x})$ corresponds to the neural fuzzy network design with appropriate weights and fuzzy membership function shapes that best match the required performance model.

While creation of a training set can be regarded as part of the preparatory overhead, this can be accomplished systematically by using a circuit simulator. Once a data set is generated and a relevant fuzzy network is trained for a given topology, the same network can be used in different synthesis applications involving the same circuit topology. In addition to modeling complex circuits that can possess substantial non-linearities, the fuzzy network approach also obviates the need for AC simulation to model performance specifications such as GBW, PM etc.

3.6 Incorporation of transistor mismatches

Design and optimization flow for modern analog integrated circuits should incorporate fabrication technology parameter variations and the resulting mismatch effects for precision and reliable operation. There is always a mismatch between two nominally identical transistors located at different coordinates across the wafer, which may cause considerable performance degradation of the overall circuit when not taken into account. The cause of mismatch can be characterized by local and global variations in parameters. The global variation accounts for the total variation in the value of the component over a wafer or a batch while the local variation reflects the variation in a component value referenced to an adjacent component on the same chip [47–48]. The local variations are random and can be modeled only by statistical approaches [49–50]. In this work, global variations are taken into consideration as the dominant cause of mismatch and they are incorporated into the analog IC optimizer in the following two steps.

- Prediction of variation in the DC operating point (ΔI_D, ΔV_{T0}, $\Delta\beta$) with respect to coordinates of the transistors in the wafer.

- Modeling the circuit-level performance variations (ΔA_o, ΔGBW, ΔSR, etc.) resulting from the transistor DC operating point variations.

The performance variations are then embedded into the main cost function of the optimizer as penalty terms to minimize the effects of the variation induced transistor mismatches on the overall performance of the integrated circuit.

3.6.1   Prediction of Mismatch

In integrated circuit fabrication, the following physical transistor parameters are dominant in causing the electrical operating point mismatches between transistors located at different coordinates of the wafer [50].

$$W, L : \text{MOS transistor channel width and length}$$
$$TOX : \text{gate-oxide thickness}$$
$$NSUB : \text{substrate doping concentration}$$
$$LD : \text{lateral diffusion length}$$
$$RSH : \text{diffusion sheet resistance}$$
$$XJ : \text{junction depth}$$

These physical parameters tend to drift away from their nominal values across the wafer due to the inherent stochastic nature of processing steps involved in integrated circuit fabrication. The nonlinear nature of transistors and the complex interaction of fabrication-related physical parameters with the transistor's circuit-level operating point typically preclude the development of complete analytical models that accurately predict mismatches. A remedy to this challenging problem is to employ neural networks in representing the mismatch relationships between physical and electrical parameters. To this end, a neural network CAD module has been developed for predicting the mismatch between transistors located across the wafer. The neural network has been trained with a data set of size 3000. The training data set has been generated via Hspice simulations by varying the above mentioned dominant physical parameters in the SPICE MOS transistor model with respect to their nominal values. The information on variation of the physical parameters across the wafer in terms of (x,y) coordinates has been gathered from typical distributions provided by semiconductor vendors [51]. The data set contains the physical transistor parameters and the corresponding electrical operating point information generated by Hspice. For all the synthesis examples, an acceptable average mismatch prediction error of 0.00012 has been found by training a multi-layer perceptron neural network with eight inputs, one hidden layer with six neurons, and one output using the back propagation technique. The trained neural network module has been capable of approximating the mismatches at a level of accuracy that can also be obtained via iterative SPICE simulations. Thus, the advantage of employing neural networks in representing mismatches is the reduction of computational load by avoiding these costly SPICE iterations.

The inputs to the mismatch prediction module are formed by the W, L parameters and x, y coordinates of each transistor on the wafer that are also duplicated for the center, resulting in a total of eight inputs. During these calculations, it is assumed that the physical parameters have their nominal values in the wafer center. Thus, the outputs of the module are drain current (I_D), threshold voltage (V_{TO}), and device transconductance parameter (β) variations of each transistor with respect to the wafer center.

3.6.2 Modeling the Circuit-Level Performance Variations

A three-layer neural-fuzzy logic network has been utilized to model the circuit performance variations arising from the variations in operating point drain currents of the transistors in the IC being optimized [46]. For the membership function, the well-known Gaussian function is used in the first layer. The second layer uses the product operation rule for fuzzy implication, and the third layer is the output defuzzifier using the center of gravity method. Hspice has been used to generate the training set according to the Latin Hypercube experimental design technique. The training set contains an output vector composed of the circuit performance variations (ΔA_o, ΔGBW, ΔSR, etc.) and an input vector of drain current variations ΔI_D. An evolutionary optimization procedure similar to the one presented in this work has been utilized in training the neural-fuzzy network. The optimization variables include fuzzy membership function parameters (width and mean values of Gaussian function) and rule numbers. The optimization technique adapts these variables while minimizing the training error based on the differences in the desired and current outputs of the neural-fuzzy logic network. The optimum number of rules and training errors of the neural-fuzzy circuit-level performance estimator for each example are detailed in the next section.

3.6.3 Incorporation of Mismatch into the Optimization System

To incorporate the mismatch model, a wafer having a maximum of $\pm$ 12.5 cm x and y coordinates with respect to the center is considered. The wafer is divided into equal segments. For representation of mismatches, the circuit topology under consideration is divided into transistor pairs such as current mirrors and differential pairs. As the optimizer adapts the transistor parameters W and L, ΔI_D is computed for every segment and a matrix $[\Delta I_D]_{n \times m}$ is formed, where n is the number of transistors and m is the number of segments in the wafer. This matrix is then input to the neural-fuzzy estimator, producing a vector $[\Delta P]_{m \times 1}$ for each performance variation. Finally, the weighted sum of each variation is embedded into the cost function in Equation 3.2 as an additional constraint defined as

$$g_j(x) = \sum_{i=1}^{m} w_{l+k+i} \, [\Delta P_j]_{i \times 1} \qquad (3.14)$$

These constraints effectively act as penalty terms for minimizing the effects of transistor mismatch. Figure 3.6 displays the interaction of the mismatch prediction module with the optimization loop.

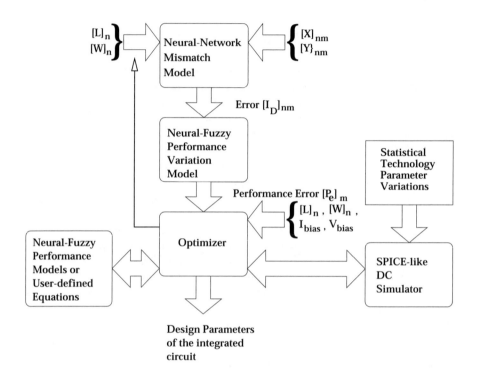

FIGURE 3.6
The mismatch prediction module.

3.7   Synthesis Examples and Discussion of Results

In this section, various analog IC synthesis examples are presented to demonstrate the validity of the proposed approach. The first example is concerned with the design of a BTS opamp. The second synthesis example deals with a second generation current conveyor circuit (CCII). The third example addresses the applicability of the presented synthesis methodology to complex topologies. To that end, a folded cascode amplifier topology has been selected and synthesized. In all the examples, the synthesis task has been driven by the circuit specifications and the results of synthesis have been validated by Hspice level 49 (BSIM3) simulation models for a 1.5 μm standard CMOS process. In addition, a prototype chip containing the synthesized designs has been fabricated and tested for further validation on silicon.

The genetic operations, empirically obtained mutation rates of strategy parameters, and selection mechanism outlined earlier regarding the optimization procedure have been applied uniformly to all the synthesis examples. The same cooling schedule has also been adopted in all the examples. In particular, an initial temperature of $T_0 = 800$ has been stepped down in a non-linear manner governed by the cooling function $T = 800(1 - g/800)^2$, where g is initially set to zero and increased by one for every ES iteration. For evaluating the performance of the circuits being optimized, both user-defined equations and neural-fuzzy performance models are used in different combinations in all the examples. The same neural-fuzzy network structure employed in modeling the mismatch induced circuit-level performance variations has also been optimized and trained for performance modeling.

The input to the automatic synthesis system is a SPICE-like netlist file containing user-defined specifications. To demonstrate the independence from initial starting point, this input file has been assigned random W and L values. The input file also contains fields for the analog cell definition pertaining to matching requirements for certain transistor pairs, DC bias voltages and currents, and the load capacitance. The examples shown below were run on a 350 MHz PC with 64 Mbyte RAM. It has been observed in all the examples that the number of cost function evaluations was more than 1000. To provide insight to the computational loads involved during synthesis runs, CPU run times are also included at the end of each example.

3.7.1   Synthesis Examples

Example 1. BTS opamp: A BTS opamp, shown in Figure 3.7, has been synthesized in this example.

For performance evaluation, both user-defined equations (for open loop gain, slew-rate, output impedance) and fuzzy performance modeling (gain-bandwidth product, phase-margin) have been used:

- - User-defined equations: The DC simulator has been employed in calculating the operating point of the circuit, subsequently yielding the small signal parameters g_m, g_o, and g_{mb}. These parameters determine the user-defined equations. The DC operating point current and voltage values have been used for power dissipation, offset voltage, and output swing calculations.

- - Fuzzy performance modeling: The training set was generated by taking 300 different base points while other variables are changed using the *Latin Hypercube* technique. For each base point, gain-bandwidth

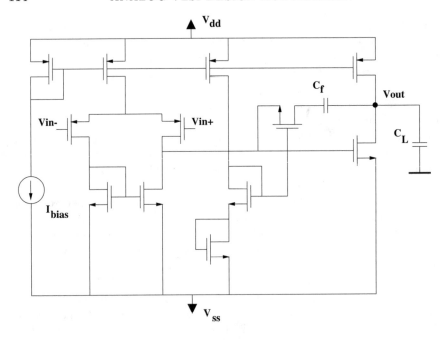

FIGURE 3.7
Schematic of the synthesized BTS opamp.

product(GBW) and phase margin (PM) have been simulated with SPICE. The optimized and trained fuzzy performance model of the gain-bandwidth product had eight rules and an average error of 0.00021 and the model for the phase-margin had eight rules and an average error of 0.00018.

Driven by the above performance evaluation models, the following cost function has been formed of constraint terms and optimized in this example.

$$C(\mathbf{x}) = w_1 GBW(\mathbf{x}) + w_2 PM(\mathbf{x}) + w_3 A_o(\mathbf{x}) + w_4 CMRR(\mathbf{x})$$
$$+ w_5 SR(\mathbf{x}) + w_6 TPWD(\mathbf{x}) + w_7 OS(\mathbf{x}) + w_8 r_{out}(\mathbf{x})$$
$$+ w_9 GBW_{mm} + w_{10} PM_{mm} + w_{11} A_{o\ mm} + w_{12} CMRR_{mm}$$
$$+ w_{13} SR_{mm} + w_{14} TPWD_{mm} + w_{15} OS_{mm} + w_{16} r_{out\ mm}$$

$$(3.15)$$

where GBW is the gain-bandwidth product, PM is the phase-margin, A_o is the DC gain, $CMRR$ is the common-mode rejection ratio, SR is the slew-rate, $TPWD$ is the DC power dissipation, OS is the output node voltage swing range, and r_{out} is the output resistance. The second group of terms in this expression pertain to the mismatch induced performance variation penalties. In this synthesis example, all the mismatch penalty terms have

TABLE 3.2 BTS Opamp Synthesis Results.

Attribute	Specification	Synthesis (Hspice level 49)
Capacitive load (pF)	1	1
Supply (V)	5	5
D.C. gain (dB)	≥ 75	81.5
Gain-bandwidth (MHz)	≥ 1	2.2
Phase margin ($^\circ$)	≥ 50	92.7
Slew rate (V/μs)	≥ 0.53	1.62
Static power (mW)	≤ 10	0.378
Output swing (V)	$\geq \pm 1.5$	± 2.48

been represented by a three-layer fuzzy-neural network trained with 10 rules. The error of the training was 0.0001. The following run times have been measured for this example: 12 min. without the mismatch module, 45 min. with the mismatch module. Synthesis and simulation results for the BTS opamp are given in Table 3.2

Example 2. CCII: A second generation current conveyor - CCII circuit as described in [52] has been synthesized in this example. Figure 3.8 shows the schematic of the circuit.

This example signifies the generality of the presented synthesis methodology when applied to non-standard circuits. In particular, user-defined equations have not been used during the optimization. For the bandwidth and mismatch performance deviation models, the previously described neural-fuzzy performance modeling tool has been used.

The search variables of the circuit comprise the bias voltage V_{bias}, compensation voltage V_c, and the MOS transistor W, L parameters. The training set was generated by taking 300 different base points, where the bias circuit shown in Figure 3.9 was used during the creation of the training data. Y input signal has been applied as a sine wave between -2 V and +2 V, while the other variables are changed using the *Latin Hypercube* technique. For each base point, V_{out} and V_x voltages have been simulated and the bandwidth has been found for both the short-circuit current and open-circuit voltage responses separately with SPICE. The optimized and trained fuzzy performance model of the open-circuit bandwidth had eight rules and an average error of 0.0014 and the model for the short-circuit bandwidth had eight rules and an average error of 0.0015. While the fuzzy performance models of the circuit have been used during the evaluation cycle of the optimization engine, the CCII has been simultaneously optimized according to the specifications.

The optimizer mainly uses the DC simulator for the offset between X-Y and Z-Y terminals. During the evaluation cycle of the optimizer, the Y terminal voltage has been swept linearly between +2 V and -2 V. The offset voltages between the terminals have been calculated by the DC simulator

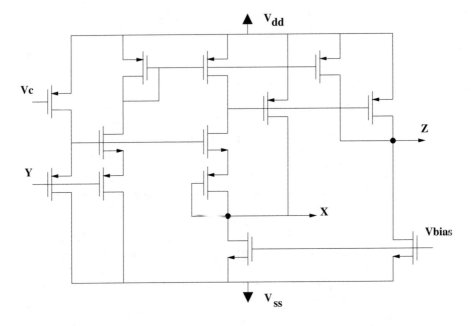

FIGURE 3.8
CCII circuit.

and the results have been treated as the inputs to the cost function of the optimizer. Hence, the synthesis system behaves in a manner to minimize the distortion while increasing the bandwidth and decreasing the terminal offsets without using expert defined equations.

The cost function for this example has been formed of both constraint and objective terms. For objective functions, the average voltage errors Avg_XY_{err} and Avg_ZY_{err} between XY and ZY terminals have been defined, respectively. The following cost function has been optimized in this example.

$$C(\mathbf{x}) = w_1 Avg\_XY_{err}(\mathbf{x}) + w_2 Avg\_ZY_{err}(\mathbf{x})$$
$$+w_3 TPWD(\mathbf{x}) + w_4 BW(\mathbf{x})$$
$$+w_5 BW_{mm} + w_6 TPWD_{mm} \qquad (3.16)$$

where BW is the bandwidth. The last group of terms in this expression pertain to the mismatch-induced performance variation penalties. In this synthesis example, all the mismatch penalty terms have been represented by a three-layer fuzzy-neural network trained with nine rules. The error of the training was 0.00012. The run times for this example is 60 min. with the mismatch module. Table 3.3 displays the synthesis and simulation results for the CCII configured as a unity gain amplifier.

Example 3. Folded cascode amplifier: The applicability of the presented synthesis methodology to complex and realistic circuit topologies is addressed

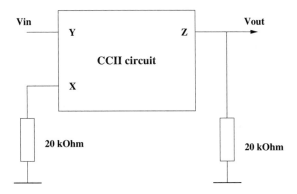

FIGURE 3.9
CCII circuit under unity gain configuration.

TABLE 3.3 CCII Unity Gain Amplifier Synthesis Results.

Attribute	Specification	Synthesis (Hspice level 49)
Supply (V)	± 5	± 5
Static power (mW)	≤ 20	11.8
XY terminal average error (mV) (between +2, -2 V input)	≤ 10	6.2
ZY terminal average error (mV) (between +2, -2 V input)	≤ 10	6.2
Bandwidth of short circuit Z terminal output current (MHz)	≤ 100	78

in this example. The circuit under consideration is a folded-cascode amplifier with common-mode feedback circuitry [9]. The schematic of the circuit is shown in Figure 3.10.

In this example, slew rate and output impedance have been modeled via user-defined equations, whereas fuzzy-neural performance models have been used for DC gain, gain-bandwidth product, phase margin, and common-mode rejection ratio (CMRR). All the fuzzy-neural models had eight rules and an average error of 0.00052 for this circuit. The offset voltage, static power dissipation, and output voltage swing values have been obtained by the DC simulator. In addition, the circuit operating point variables such as g_m, g_o, and I_D have been computed by the DC simulator in order to form the user-defined equations.

The cost function of this example is similar to that of first example, with

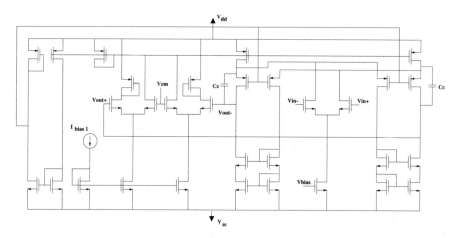

FIGURE 3.10
Folded cascode topology with common-mode feedback.

the exception of GBW being an objective term that needs to be maximized. The cost function is

$$C(\mathbf{x}) = w_1/GBW(\mathbf{x}) + w_2 PM(\mathbf{x}) + w_3 A_o(\mathbf{x}) + w_4 CMRR(\mathbf{x})$$
$$+ w_5 SR(\mathbf{x}) + w_6 TPWD(\mathbf{x}) + w_7 OS(\mathbf{x}) + w_8 r_{out}(\mathbf{x})$$
$$+ w_9 GBW_{mm} + w_{10} PM_{mm} + w_{11} A_{o\ mm} + w_{12} CMRR_{mm}$$
$$+ w_{13} SR_{mm} + w_{14} TPWD_{mm} + w_{15} OS_{mm} + w_{16} r_{out\ mm}$$

$$(3.17)$$

In this synthesis example, all the mismatch penalty terms have been represented by a three-layer fuzzy-neural network trained with eight rules. The error of the training was 0.00015. The following run times have been measured for this example: 45 min. without the mismatch module, 124 min. with the mismatch module. Table 3.4 displays the synthesis and simulation results for the folded cascode topology.

3.7.2　Validation on Silicon

A prototype test chip incorporating the automatically synthesized BTS opamp, CCII, and folded cascode circuits has been fabricated with a 1.5 μm standard CMOS technology. Figure 3.11 displays the micrograph of the test chip.

Various tests and measurements have been performed on the test chip in order to validate the performance of the presented synthesis strategy on silicon. The following observations have been made.

TABLE 3.4 Folded Cascode Amplifier Synthesis Results.

Attribute	Specification	Synthesis (Hspice level 49)
Supply (V)	$\pm$ 5	$\pm$ 5
Static power (mW)	$\leq$ 30	7.59
D.C. gain (dB)	$\geq$ 60	65.4
Gain-bandwidth (MHz)	$\uparrow$	94
CMRR	$\geq$ 100	765
Phase margin ($^\circ$)	$\geq$ 50	88
Output impedance (kΩ)	$\geq$ 80	100
Offset voltage (V)	$\leq$ 0.4	0.2
Output swing (V)	$\geq \pm$ 1	$\pm$ 1.6

- For the BTS opamp, the D.C gain has been measured as 76 dB, the gain-bandwidth product has been observed to be 1.13 MHz, and the phase margin has been measured as 84°. The large signal performance of the opamp has been characterized by a slew rate of 0.64 V/μs and an output swing range of $\pm$ 2.3 V.

- For the CCII, the Y terminal voltage has been connected to a sinusoidal voltage source and the X and Z terminal voltages have been observed under the unity-gain configuration as depicted in Figure 3.9. To demonstrate the current conveyor principle, XY terminal voltage pair swept between +2 V and -2V is displayed in Figure 3.12.

- For the folded cascode topology, the common-mode feedback circuit operation has been verified by open-loop operation. Since this topology is a fully differential one with a substantially high gain-bandwidth, the measurement setup has not been reliable for determining the unity gain frequency. However, the open-loop amplifier operation has been easily observed at high frequencies due to dominant pole behavior. Figure 3.13 displays the time-domain response of the amplifier for an input frequency exceeding 200 kHz, where the upper waveform represents the output.

- The discrepancies between Hspice and measurement results are mainly due to errors in reliably estimating the capacitive loading conditions (especially for AC specifications).

3.7.3 Discussion of Results

For all the three examples above, simulation results show excellent agreement with synthesis objectives, remaining well below the criteria. It is generally the case with circuit optimizers that the resulting designs are at the edge of the design space in that a small deviation in one of the parameters will cause a big variation at the output. This is not the case for our system

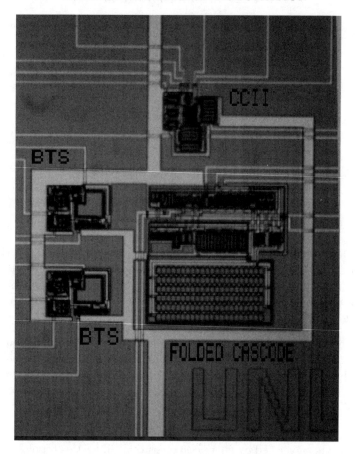

FIGURE 3.11
Micrograph of the test chip (after [53], © 2003 IEEE).

where the results from silicon still obey the design criteria, especially for DC parameters. The total DC power for all the three circuits was measured to be 28 mW including the power dissipation in 40 pads of the test chip. The simulated total power dissipation for all the circuits is around 20 mW. Unfortunately, the power excluding the pads cannot be measured, but it is safe to state that the DC conditions are met in the designs. This is even clearer from the BTS opamp gain, which was simulated to be 81.5 dB and measured to be 76 dB or from the output swing whose simulated and measured values are 2.48 V and 2.3 V, respectively. Unfortunately, the same things cannot be stated for the AC characteristics. Although the measured values still remain within the specified bounds, they are far from simulated values. However, this does not show a weakness in our system. On the contrary, it shows that the SPICE parameters of the fabricated design are substantially different from the

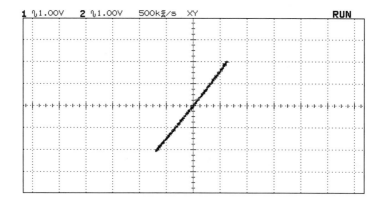

FIGURE 3.12
CCII *XY* terminal pair swept between +2 V and -2 V (after [53], © 2003 IEEE).

previously supplied SPICE parameters of the manufacturer. New simulations were done with the parameters supplied after fabrication and the simulation results were much closer to measured results. It can be stated that this is another strength of our system because the fabricated circuit still remains within specifications even though the process varies greatly from run to run. Although our main aim was to circumvent the effects of variation within a wafer, the inclusion of parameter variations within the optimization loop resulted in a more robust design with respect to inter-wafer variations. Similar comments can be made about the other two examples. It should be noted that the three examples are very different from each other in topology and in design objectives. These examples demonstrate the ability of the presented approach to synthesize a broad range of analog integrated circuits. Although the presented results are based on a $1.5\mu m$ CMOS technology due to the relatively low cost of prototype fabrication, state-of-the-art technologies with smaller transistor sizes can also be supported in this synthesis scheme, since a SPICE-like simulator runs inside the optimizer.

3.8 Concluding Remarks

Circuit-level synthesis currently remains as one of the most critical tasks required for full automation of analog VLSI systems. Thus, the past two decades have witnessed substantial progress in this area with contributors from both industry and academia. We have provided a literature survey in the

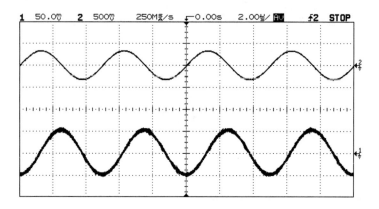

FIGURE 3.13
Time-domain response of the folded cascode amplifier (after [53],
© 2003 IEEE).

beginning of this chapter on this progress. While this survey can be regarded
as concise, the interested reader can use the taxonomies provided herein for
developing a perspective and pursuing his/her needs in this field. A major
outcome of the survey on circuit-level synthesis systems is the need for a tool
that can synthesize a broad class of analog circuits with manufacturability
and reliability constraints. Furthermore, the expert design knowledge should
not be a strict prerequisite for being able to effectively use these tools. In
our opinion, only a tool with these compliances can be a widely accepted one
within the designer community.

While there are promising developments toward establishing complete analog circuit synthesis environments, there still exists considerable challenges.
These can be listed as

- inclusion of efficient simulators in the evaluation cycles,

- very fast exploration of search variable space,

- minimizing user-defined equations for keeping the preparatory overhead
  low and manageable,

- handling non-standard topologies,

- incorporation of variation-tolerance and reliability constraints,

- easy migration to different technologies.

We attempted to address these issues to an extent by developing an analog
synthesis system. To that end, the application of evolutionary strategies to

analog integrated circuit optimization problem has been discussed. A novel optimization algorithm based on a combination of ES and simulated annealing has been developed. This is realized by borrowing the stochastic Metropolis criterion from SA and using it as the selection mechanism of ES in finding a satisfactory solution. A SPICE-like simulator has also been developed for DC parameter optimization and neural-fuzzy and/or equation-based models have been devised for AC parameter optimization. Parameter variation effects due to manufacturing tolerances have also been included in a novel manner in the optimization loop, thus making the end result more robust with respect to variations and mismatches.

The overall synthesis system exists as a set of interconnected software modules. As in every analog design automation environment, there is a level of preparatory effort and design experience involved in our system prior to synthesis runs. This includes the generation of user-defined equations, training the neural network for performance models and incorporation of mismatches. By developing the necessary software interface modules, the presented system can also be a third party module in an existing industrial design automation environment.

Our approach has been demonstrated on several examples, and convergence to the desired performance criteria has been attained rather easily. The computation times of examples have shown that circuits of moderate complexity can be synthesized in reasonable amounts of time by avoiding expensive simulation iterations. This has been made possible by employing fast neural models and limiting the simulations to the DC domain only for speed-up. The synthesized designs have also been simulated and verified with Hspice using the industry standard level-49 transistor simulation models. The simulations show that the circuits designed by our tool conform to the synthesis specifications. A prototype chip containing the synthesized circuits has been manufactured and measured. The measurements demonstrate the validity of the presented evolutionary approach to automatic synthesis of analog integrated circuits.

Finally, the structure of the circuit-level optimizer is such that it permits the formation of a complete analog design automation environment by interfacing the presented system with automatic topology selection residing at a higher level and physical layout generation modules residing at a lower level.

References

[1] Gielen, G. and Rutenbar, R.A., Computer-aided design of analog and mixed-signal integrated circuits, *Proceedings of the IEEE*, 1825, 88, 2000.

[2] Degrauwe, M.G.R., Olivier, N., and Dijkstra, E., IDAC: An interactive design tool for analog CMOS circuits, *IEEE J. Solid-State Circuits*, 1106, 22, 1987.

[3] Harjani, R., Rutenbar, R.A., and Carley, L.R., OASYS: A framework for analog synthesis, *IEEE TCAD*, 1247, 8, 1989.

[4] Sheu, B.J., Fung, A.H., and Lai, Y., A knowledge-based approach to analog IC design, *IEEE Trans. Circuits Syst.*, 256, 35, 1988.

[5] El-Turky, F. and Perry, E.E., BLADES: an artificial intelligence approach to analog circuit design, *IEEE TCAD*, 680, 8, 1989.

[6] Koh, H.Y., Sequin, C.H., and Gray, P.R., OPASYN: A Compiler for MOS operational amplifiers, *IEEE TCAD*, 113, 9, 1990.

[7] Nye, W., Riley, D.C., Sangiovanni-Vincentelli, A., and Tits, A.L., DELIGHT.SPICE: An optimization-based system for the design of integrated circuits, *IEEE TCAD*, 501, 7, 1988.

[8] Gielen, G.E. and Sansen, W.M.C., Analog circuit design optimization based on symbolic simulation and simulated annealing, *IEEE J. Solid-State Circuits*, 707, 25, 1990.

[9] Ochotta, E.S., Rutenbar, R.A., and Carley, L.R., Synthesis of high-performance analog circuits in ASTRX/OBLX, *IEEE TCAD*, 273, 15, 1996.

[10] Fares, M. and Kaminska, B., FPAD: A fuzzy nonlinear programming approach to analog circuit design, *IEEE TCAD*, 785, 14, 1995.

[11] Makris, C.A. and Toumazou, C., Analog IC design automation: Part II - automated circuit correction by qualitative reasoning, *IEEE TCAD*, 239, 14, 1995.

[12] Medeiro, F., et al., A statistical optimization-based approach for automated sizing of analog cells, *Proc. ACM/IEEE Int. Conf. Computer-Aided Design (ICCAD)*, 594, 1994.

[13] Phelps, R., Krasnicki, M., Rutenbar, R.A., and Carley, L.R., Anaconda: Robust synthesis of analog circuits via stochastic pattern search, *Proc. Design Automation Conference*, 1999.

[14] Krasnicki, M., Phelps, R., Rutenbar, R.A., and Carley, L.R., Maelstrom: Efficient simulation-based synthesis for custom analog cells, *Proc. Design Automation Conference*, 1999.

[15] Onodera, H., Kanbara, H., and Tamaru, K., Operational-amplifier compilation with performance optimization, *IEEE J. Solid-State Circuits*, 466, 25, 1990.

[16] Harvey, J., Elmasry, M., and Leung, B., STAIC: An interactive framework for synthesizing CMOS and BiCMOS analog circuits, *IEEE TCAD*, 1402, 11, 1992.

[17] Maulik, P.C., Carley, L.R., and Rutenbar, R.A., A mixed integer nonlinear programming approach to analog circuit synthesis, *Proc. 29th ACM/IEEE Design Auto. Conf.*, 698, 1992.

[18] Van der Plas, G., et al., AMGIE-A synthesis environment for CMOS analog integrated circuits, *IEEE TCAD*, 1037, 20, 2001.

[19] Hershenson, M., Boyd, S., and Lee, T., GPCAD: A tool for CMOS op-amp synthesis, *Proc. IEEE/ACM Int. Conf. Computer-Aided Design (ICCAD)*, 296, 1998.

[20] Hershenson, M., Boyd, S., and Lee, T., Optimal design of a CMOS op-amp via geometric programming, *IEEE TCAD*, 1, 20, 2001.

[21] Medeiro, F., Perez-Verdu, B., Rodriguez-Vazquez, A., and Huertas, J., A vertically integrated tool for automated design of $\Sigma - \Delta$ modulators, *IEEE J. Solid-State Circuits*, 762, 30, 1995.

[22] Doboli, A., Nunez-Aldana, A., Dhanwada, N., Ganesan, S., and Vemuri, R., Behavioral synthesis of analog systems using two-layered design space exploration, *Proc. ACM/IEEE Design Automation Conf. (DAC)*, 951, 1991.

[23] Antreich, K. and Koblitz, R., Design centering by yield prediction, *IEEE TCAD*, 88, 29, 1982.

[24] Director, S.W., Feldmann, P., and Krishna, K., Optimization of parametric yield: A tutorial, *Proc. IEEE Custom Integrated Circuits Conf.*, 3.1/1, 1992.

[25] Mukherjee, T., Carley, L.R., and Rutenbar, R., Synthesis of manufacturable analog circuits, *Proc. ACM/IEEE Int. Conf. Computer-aided Design (ICCAD)*, 586, 1995.

[26] Debyser, G. and Gielen, G., Efficient analog circuit synthesis with simultaneous yield and robustness optimization, *Proc. IEEE/ACM Int. Conf. Computer-Aided Design (ICCAD)*, 308, 1998.

[27] Schwenker, R., Schenkel, F., Graeb, H., and Antreich, K., The generalized boundary curve—A common method for automatic nominal design

and design centering of analog circuits, *Proc. IEEE Design Automation and Test in Europe Conf. (DATE)*, 42, 2000.

[28] Antreich, K., Graeb, H., and Wieser, C., Circuit analysis and optimization driven by worst-case distances, *IEEE TCAD*, 57, 13, 1994.

[29] Antreich, K., et al., WiCkeD: Analog circuit synthesis incorporating mismatch, *Proc. IEEE Custom Integrated Circuits Conf.*, 2000.

[30] Horrocks D.H. and Khalifa, Y.M.A, Genetically derived filters using preferred value components, *Proc. IEE Colloq. on Linear Analogue Circuits and Systems*, 1994.

[31] Grimbleby, J.B., Automatic analogue network synthesis using genetic algorithms, *Proc. First Int. Conf. Genetic Algorithms in Engineering Systems: Innovations and Applications (GALESIA)*, 53, 1995.

[32] Zebulum, R.S., Pacheco, M.A., and Vellasco, M., Comparison of different evolutionary methodologies applied to electronic filter design, *1998 IEEE Int. Conf. on Evolutionary Computation*, Piscataway, NJ: IEEE Press, 1998, 434.

[33] Koza, J.R., Bennett, F.H., Andre, D., Keane, M.A., and Dunlap, F., Automated synthesis of analog electrical circuits by means of genetic programming, *IEEE Trans. on Evol. Comput.*, 109, 1, 1997.

[34] Lohn J.D. and Colombano, S.P., Automated analog circuit synthesis using a linear representation, In M. Sipper, D. Mange, and A. Pérez-Uribe, editors, in *Proc. Second Intl. Conf. on Evolvable Systems: From Biology to Hardware (ICES98)*, Lecture Notes in Computer Science, Springer-Verlag, Heidelberg, 1998, 125.

[35] Zebulum, R.S., Sinohara, H.T., Vellasco, M.M.R., Santini, C.C., Pacheco, M.A.C., and Szwarcman, M.H., A Reconfigurable platform for the automatic synthesis of analog circuits, *Proc. Second NASA/DoD Workshop on Evolvable Hardware*, 91, 2000.

[36] Stoica, A., Keymeulen, D., Zebulum, R., Thakoor, A., Daud, T., Klimeck, G., Jin, Y., Tawel, R., and Duong, V., Evolution of analog circuits on Field Programmable Transistor Arrays, *Proc. Second NASA/DoD Workshop on Evolvable Hardware*, 99, 2000.

[37] Kruiskamp, W. and Leenaerts, D., Darwin: Cmos opamp synthesis by means of a genetic algorithm, *Proc. Design Automation Conference*, 1995.

[38] Sripramong, T.and Toumazou, C., The invention of CMOS amplifiers using genetic programming and current-flow analysis, *IEEE Trans. on Computer-Aided Design of Integrated Circuits and Systems*, 1237, 21, 2002.

[39] Alpaydın, G., Erten, G., Balkır, S., and Dündar, G., Multi-level Optimization Approach to Switched-Capacitor Filter Synthesis, *IEE Proceedings – Circuits, Devices, and Systems*, 243, 147, 2000.

[40] Vladimirescu, A., *The Spice Book*, John Wiley & Sons, Chichester, New York, 1993.

[41] Bayraktaroğlu, İ., Öğrenci, S., Dündar, G., and Balkır, S., ANNSyS: An Analog Neural Network Synthesis System, *Neural Networks*, 325, 12, 1999.

[42] Back D. and Hoffmeister, F., Extended selection methods for genetic algorithms, *Proc. 4th International Conference on Genetic Algorithms*, 92, 1991.

[43] Bäck, T. and Schwefel, H.P., Evolution strategies I: Variants and their computational implementation, *Genetic Algorithms in Engineering and Computer Science*, G. Winter, J. Perieaux, M. Gala and P. Cuesta, Eds. Chichester: Wiley, 1995, 111.

[44] Beyer, H-G., *The Theory of Evolution Strategies*, Springer-Verlag, Berlin, 2001.

[45] Metropolis, N., Rosenbluth, A.W., Rosenbluth, M.N., and Teller, A.H., Equation of state calculations by fast computer machines, *J. Chemical Physics*, 1087, 21, 1953.

[46] Torralba, A., Chavez, J., and Franguelo, L.G., Circuit performance modeling by means of fuzzy logic, *IEEE Trans. on Computer-Aided Design of Integrated Circuits and Systems*, 1391, 15, 1996.

[47] Pelgrom, M.J., Duinmaijer, A.C., and Welbers, A.P., Matching properties of MOS transistors, *IEEE J. Solid-State Circuits*, 1433, 24, 1989.

[48] Lakshmikumar, K.R., Hadaway, R.A., and Copeland, M.A., Characterization and modeling of mismatch in MOS transistors for precision analog design, *IEEE J. Solid-State Circuits*, 1057, SC-21, 1986.

[49] Debyser, G., Gielen, G., and Sansen, W., Efficient statistical analog IC design using symbolic methods, *Proc. International Symposium on Circuits and Systems*, 50, 2, 1998.

[50] Swidzinski, J.F., Styblinski, M.A., and Xu, G., Statistical behavioral modeling of integrated circuits, *Proc. International Symposium on Circuits and Systems*, 231, 2, 1998.

[51] Tulunay, G., Dundar, G., and Ataman, A., A new approach to modeling statistical variations in MOS transistors, *Proc. International Symposium on Circuits and Systems*, 757, I, 2002.

[52] Elwan, H.O. and Soliman, A.M., A novel CMOS current conveyor re-
     alization with an electronically tunable current mode filter suitable for
     VLSI, *IEEE Transactions on Circuits and Systems*, 663, 43, 1996.

[53] Alpaydin, G., Balkır, S., and Dündar, G., An evolutionary approach
     to automatic synthesis of high-performance analog integrated circuits,
     IEEE Transactions on Evolutionary Computation, 7, 3, 2003.

Chapter 4

Layout-Level Design Automation

4.1 Introduction

The lowest level of design hierarchy in the design automation system proposed in this book is the layout level. This level should take in a circuit description in some format at the transistor level and should output a layout description in some other format. However, the problem is not a simple geometrical one since the layout directly affects the performance of the circuit through several criteria, the most important being parasitics and device matching. In the subsequent pages, fundamentals of layout design automation and the specific problems pertaining to analog design will be presented along with a comprehensive review of the literature in this area.

The work on analog layout automation started very early on and this area is one of the most studied areas in analog design automation besides circuit-level automation. One of the first attempts at layout automation [1] makes the simple assumption that manual schematics have the best topology for the layout and simply replaces the components from the circuit diagram with the associated layouts created parametrically. A similar approach is also employed in [2]. However, the user can annotate relationships on the schematic in this case. These relationships are taken into account during the placement phase of the generated devices. After placement, routing is performed via the conventional maze routing algorithm. If complete routing cannot be achieved, rip-up and reroute can be applied manually. Obviously, these approaches are a long way from being automatic, as they rely on the initial circuit schematic, which is not necessarily the best topology and which may even not be present in the first place. In [3], the topology is learned from an "example." For a given module, an expert designer draws a layout that is believed to be optimal. The topology is extracted from that layout and the layout generator uses it in all subsequent designs. This way, the expert's knowledge in the areas of device placement, wire trajectories, and positions of module placements is captured. In [4], a similar idea is utilized. Here, the software, OO-ADS, uses a set of floorplans captured by an expert designer beforehand. The actual designer chooses among these floorplans for the current design. However, one should note here that an optimal topology for a set of device dimensions may not be the optimum one for another set. The optimal topology may even vary for the same sized circuit depending on the application and performance parameters of interest. Nevertheless, such an unsophisticated approach is still not useless. For example, it can be used for retargetable layout as explained in [5]. CMOS technology shrinks continuously and, with each new technology, circuits have to be re-optimized and layouts must be re-designed according to slightly changed sizes and design rules. It is conceivable that optimum topologies will not change that much with such minor changes from technology to technology, thus making predefined topologies a good choice.

Another independent approach to analog layout automation has been inspired by digital layout automation [6–8]. The modules are designed as standard cells and placed in a standard cell arrangement, which is not really suitable for analog design. However, [8] presents one of the earliest mixed-signal layouts generated automatically.

One of the major areas in which early analog layout automation tools have had big problems was determining the topology of the layout. Some researchers have tried to solve this problem by defining a library of topologies and using this library for layout generation. Borutzky et. al. [9] define a list of building blocks that they refer to as "skeletons." These skeletons have static geometrical relationships within themselves and can be sized according to the circuit design. Furthermore, skeletons can be hierarchical within themselves. The whole design can be obtained by combining these predesigned parametric skeletons. The placement of the skeletons is achieved by three cooperating expert systems; namely, the global placer, the local placer, and the constraint equalizer. A gridless Lee router completes the layout. Another tool working on the same skeleton idea was developed independently by Bowman [10]. He calls these building blocks object-oriented imaging models and compares them to letter definitions in a laser printer where each letter is defined in terms of lines and arcs and their relative positions. These definitions are used to form the letters, whatever the size of the font. His macrocells include resistors, capacitors, transistors, differential pairs, current mirrors, and differential pair plus active load.

In digital standard cells, the topology of each cell is more or less fixed. However, the same cannot be said for analog circuits because the topology will change for the same circuit diagram when device sizes vary. Therefore, the above approaches where the layout is static is not very general and cannot be expected to yield very good results. In [11], the geometric topology definition is left to the designer. The layout-level design automation tool is part of a CMOS operational-amplifier compiler OAC, which also carries out the device sizing. Once the device sizing is finished, the user gives instructions to the layout by utilizing a layout definition language with which the layout is described in terms of component locations, combinations of components, orientation of components, etc. Obviously, this is not the solution for an automatic layout generation, which the authors claim is too difficult and requires a lengthy computation time. Such instructions are given to the layout generator in STAIC [12], [13] as well, but it is the circuit-level optimizer and the library that generate these instructions. As discussed in the previous chapter, STAIC is a design automation tool that works at the circuit level and at the layout level. The layout-level design automation tool of STAIC is barely more than a module generator, however. The information about the topology is passed from the library and the circuit level. Thus, the layout description is created from a solved geometry. The place and route directives are embedded in the selected layout style descriptions.

As opposed to the class of algorithms where the layout is static, automatic layout generators that can perform topological arrangement with flexible layout blocks became available very early on as well. One such tool was presented in [14]. The tool consists of a small set of parameterized leaf cell generators that are used to create the layout for simple components. A floorplan is created from these cells by using slicing trees. The routing is done via a switchbox router, MIGHTY [15], already available to the authors. After layout is completed, compaction is applied. This approach remains as a valid approach even today if one measures the success of the layout only geometrically.

However, one of the major difficulties in analog layout is to optimize the layout not only in terms of geometry, but also in terms of performance. The authors are aware of this fact and have tried to incorporate some performance parameters into their algorithm. For example, nets are assigned to a few categories with different priorities. Top priority nets are routed first, thus minimizing the possibility of crosstalk. However, their approach does not go beyond some heuristics. ALDA [16] and its newer version, ALDA2 [17] are tools that process the incoming circuit in three steps; partitioning, floorplanning, and physical assembly. Partitioning is done in a scoring-based manner, whereas a slicing structure is used for floorplanning. Physical assembly is not carried out by the tool itself, it rather controls a standard CAD framework for layout design. A later version of ALDA [18] carries this idea further and uses two data interface formats. One format is called the "hard data interface" and drives the commercial CAD tool. The "soft data interface" provides reduced geometric data, which is sufficient for partitioning and floorplanning. This version of ALDA is also able to recognize specific structures from the netlist such as differential pairs, cascoded or simple current mirror clusters, series or parallel connected transistor clusters, and cascode transistor clusters. Furthermore, it generates the corresponding layouts for these blocks.

Another early example of automatic layout generators is SALIM [19], [20]. M. Kayal et al. again propose the parameterized creation of leaf cells and an initial placement via a force-directed algorithm. The floor-plan generation is performed in a bottom-up manner. A strong point of their tool is that the leaf cells can be created in different shapes, thus helping in the floorplanning and placement phase. They also propose that the user should be able to intervene in the layout generation process at any time and change the layout. Thus, their tool is interactive. This interactive layout generation idea was pursued further in creating ILAC [21], [22] which is still one of the best-known tools for layout automation. ILAC is a companion tool for IDAC, which was discussed in the previous chapter. In ILAC, which is actually a suite of tools, the circuit is partitioned into functional blocks where the tool STUCCO provides shape and interconnect information for all possible realizations of a functional block as discussed in [19] and [20]. MOSAIC is the name of the tool used for placement and interconnection. Simulated annealing is utilized for floorplanning and global routing whereas the channel router is based on scanline techniques and supports rip-up and reroute. A slicing structure is used for placement, but a further optimization is performed via simulated annealing. Both MOSAIC and STUCCO are technology independent. A final clean-up run is done by PICTURE, which postprocesses the layout and maps it to the technology. Again, ILAC allows user intervention at all levels and is truly interactive at all levels. However, it is still not performance based. Interactive tools, however, have not fallen out of favor in subsequent years, as indicated by the fact that new tools, such as the tool presented in [23], have appeared in the literature. Their advantage is obvious in that the difficult problem of performance optimization is left to the layout designer. In many cases, the layout designer prefers to obtain these layouts by hand rather than rely on the generation software.

Another method of obtaining expert information without having the user intervene at all levels is being able to recognize building blocks when faced with a circuit netlist, similar to the approach in ALDA. ALSYN [24], [25] was one of the first tools with this capability. The circuit whose layout is to be generated is first analyzed by a compiled rule set. Basic building blocks such as current mirrors are identified and collected into a "bag." Rules exist at all levels of hierarchy, recognizing large structures. The placement program, PALM, determines a slicing structure using a MinCut approach. Area optimization is carried out by selecting the most suitable variant of each module among all available. Routing is achieved via a simple maze router.

The first hints of performance parameters appearing in analog layout automation can be traced back to ANAGRAM [26] and LADIES [27], both of which have become quite well known in the analog design automation community. ANAGRAM is a tool consisting of a module generator, a placer based on simulated annealing, and a line expansion router. The router is tailored for performance optimization. Nodes are classified into three types; namely, neutral, noisy, and sensitive nodes. The router is given a penalty for crosstalk depending on the closeness or overlap of certain pairings like noisy and sensitive nodes. Thus, the router has become performance-oriented. However, the classification of nodes has to be done by hand before ANAGRAM is run. LADIES is actually a suite of tools, some of which try to interpret the performance information, which is converted to constraints like concatenating devices, placing them near or apart from each other, etc. These constraints are generated either by a human or by LISA, which is a knowledge-based system that analyzes the circuit netlist. SOPHIA is an analog placer based on the concept of clustering and ROSA does the routing in a multilayer gridless fashion. Finally, PRIMA takes in the constraint information along with the placed and routed layout and applies the constraint information. With this information, PRIMA both compacts the design and corrects the layout where it conflicts with constraints. Neither ANAGRAM nor LADIES is a completely performance-oriented tool in our opinion, but it can be said that they are "partially performance-aware tools." The ANAGRAM tool was later improved and updated [28], [29]. The new ANAGRAM was restricted to routing only, but several improvements were done on the routing algorithm such as allowing arbitrary wire pitch over the device wiring, and rip-up and reroute during routing even when not necessary. This new router was called ANA-GRAM II. The placement task was allocated to a new tool, KOAN. KOAN has several advantages over ANAGRAM in that it allows placement of symmetrical objects, makes internal geometry of all cells visible so that merging can be exploited, and automatically generates wells and substrate contacts. In later years, KOAN/ANAGRAM II has arguably become a standard for analog layout automation. INALSYS [30] employs a methodology for module generators, which is between having a large library of predefined modules (like in ILAC, ALSYN, or SALIM) and having almost no predefined modules (like in KOAN). INALSYS has a compact set of parameterizable modules. Floorplanning is done based on a simulated annealing optimized slicing tree model. Here, three costs are defined; namely, area, net length, and well merging. Routing is carried out in two phases, global and detailed routing. An improved switchbox model is used for detailed routing.

The first performance-oriented tools were presented in [31], [32], and [33]. In [31], the circuit netlist is analyzed based on a rule-based system, which classifies the nodes into various categories and recognizes basic structures. Then, based on this information, critical nodes are identified and a sensitivity analysis is done on these nodes. Analog placement is performed on parameterized modules via a slicing tree based on sensitivities as well as geometric considerations. Routing is carried out via a MIGHTY-type router. The final phase is the performance-driven optimization phase where reliability information is also included while performing compaction. Hong and Allen [32], [33] also perform a sensitivity analysis on the circuit before the actual layout phase. During the sensitivity analysis, extra components modeling the parasitics originating from interconnections are inserted to the circuit. Placement is done based on a partitioning scheme. All possible partitions are generated and enumerated based on the performance information. The best partition forms a basis for subsequent placement. Routing is carried out using MIGHTY. The layout automation system naturally includes device generators as well. In [34], the constraints are generated through simulation, with the harmonic balance method first. Then, automatic placement is performed via a simulated annealing-based constraint-driven program. It is demonstrated in this work that minimizing only area during placement may lead to very unfavorable routing situations. Finally, after completion of placement, constraints are generated for routing, which is carried out with MIGHTY with a priority functionality. A formal definition for performance constraints and an automation methodology using them was presented in [35]. This work has become a reference for performance-oriented layout generation over the years. The authors describe many important concepts in calculating performance constraints for layout design and using these in the actual layout-generation phase. One of these concepts is the "sensitivity matrix." This is a matrix quantifying the sensitivity of each performance parameter of the circuit to each layout effect. The layout effects can be quantified as either parasitics or device matching constraints. This issue will be discussed later in this chapter in more detail. From this matrix, cost figures, which are used during optimization, are derived. The costs used in driving the layout generation are total wire length, area, overlap, discontinuities in wells, distance between placement and symmetric configuration, mismatches, and performance constraint violation penalties. Dynamic abutment technique is used during layout generation. This way, both overall area and parasitic junction capacitances are reduced. All of these placement and abutment operations are carried out using PUPPY-A. Transistor stacks are generated using LDD. These tools will also be discussed later on in this chapter. Routing is defined as having two distinct types, channel routing and area routing. The channel router (ART) is a two-layer gridless router, whereas the area router ROAD is an improved version of a classical maze router. Finally, compaction is performed on the circuit via SPARCS-A, which is a mono-dimensional constraint graph-based compactor.

Chen and Sheu [36], [37] combine the idea of module recognition with constraint-driven layout synthesis. Although their approach is not fully performance oriented, it is still a very interesting example. Given a circuit netlist, their software is able to recognize special sections. These sections are generated by the use of module generators. A critical net analysis is also performed on the circuit, based on the recognition, and nets are divided into seven different categories. Routing is performed by MIGHTY based on these priorities. Thus, performance constraints are partially taken into account both during the creation of modules and routing.

Charbon et. al. [38] argue that using performance criteria as explained above is not enough to meet tight specifications. The conventional approach of doing module generation, placement and routing sequentially will simply not yield enough performance. According to them, performance-oriented simultaneous module generation and placement is required for better layouts. They also argue that pre-constructed modules are a limitation. They present a model consisting of a dynamic programming model and a clique problem by which simultaneous place and route can be achieved. Similarly, the authors in [39] argue that sequential module generation, placement and routing yields sub-optimal results. However, they are in favor of automatic module recognition and generation of these modules through module generation. They advocate simultaneous place and route as the key to success in their tool RACHANA. Cohn et. al. also propose simultaneous place and route by stating that any conventional tool like KOAN/ANAGRAM II can be modified to this end simply by defining wires as blocks for placement similar to components of modules [40]. Prieto et. al. also advocate simultaneous place and route [41], [42] in their tool GELSA. They use slicing structures and simulated annealing for floorplanning and placement. Symmetry information is also considered. In fact, they define two types of symmetries; global symmetry with respect to a set of virtual axes and local symmetry affecting groups of cells. In addition, they also show how sensitivity information as obtained in [35] can be incorporated into their tool. Kubo et. al. [43] apply the idea of treating wires as modules for simultaneous place and route optimization. However, for placement, they use the sequence-pair algorithm, which will be described in the later sections.

In presenting their tool, LAKE, the authors in [44] agree with the fact that sequential place and route will not yield the best results in that the interaction between them is not controlled, especially if the only cost for placement is net length. However, they disagree with the assertion above that simultaneous place and route is the solution to this problem. According to them, simultaneous place and route is too complex, requires too much computation time, and still yields sub-optimal results because of the complexity of the problem. Furthermore, they also state that module recognition from a predefined library of structures is not general. Their system performs a pre-layout analysis and tries to extract performance-related structures from the circuit netlist according to a set of rules. These extracted sections are called slots. Simulated evolution is used for slot placement. After placement is complete, pin assignment, feed-through assignment, analog routing with priorities assigned to symmetrical structures and post cell trimming are carried out.

The extensive survey of the literature above demonstrates that there is no single "correct" approach to the design automation of analog integrated circuit layouts. The approaches range from fully interactive to fully automatic approaches with many variants in between. A fully interactive approach boils down to a simple module-generation software controlled by the designer and is already available as extensions to professional layout drawing software. A fully automatic approach has been observed to arouse suspicion and distrust from experienced designers due to the many criteria that have to be taken into account during layout generation. In our current tool, ALGv2 (Automatic Layout Generator), we have opted for automated generation, but with performance directives to be given prior to generation. These are simple directives, which can be provided by the user or generated automatically by another software that will be discussed later in this chapter. Furthermore, the user has the opportunity to "commit" the layout after each phase and make changes on it if he/she so desires.

In the next sections, various phases of automatic layout generation will be discussed along with sample results from ALGv2. Section 4.2 will describe module generation, whereas Section 4.3 will concentrate on partitioning. A subject that is very closely related to partitioning, namely floorplanning, will also be the subject of Section 4.3. Sections 4.4 and 4.5 will cover placement and routing, respectively. Section 4.6 will deal with post-layout improvement methods, the most important of which is compaction. Section 4.7 will discuss performance issues in analog layout generation and our associated tool, the layout advisor. Section 4.8 will conclude this chapter with some thoughts about the future direction of analog layout generation.

4.2   Device Generation

The device generation problem is simply the problem of generating a device, or a group of closely related devices, according to the required geometry or specifications. It can be as simple as generating the layout of a single rectangular transistor, given the dimensions, or as complex as generating a set of matched or cascaded transistors with certain electrical parameters. All modern commercial layout tools contain some kind of device-generation option, but these are unfortunately rather primitive in most cases. In this section, various device generators from the literature will be discussed along with the device generator in ALGv2. Results from ALGv2 will also be presented,

The issue of device generation began to attract attention by itself in the early 1990s. One of the earliest studies in this area [45] focused on the generation of binary weighted capacitor arrays and scaled switches. Both structures are very important for A/D and D/A converters and a bad layout can cause important errors, as described in Chapter 2. The device generators take as inputs electrical goals and area goals and generate possible geometrical dimensions and aspect ratios. Later, they generate the mask layouts from the selected optimal geometries. In a later work [46], the authors have added a router to complete the design of a D/A converter with these generated devices. The generation of capacitor arrays attracted much attention during the following years. One very recent work [47] focuses on this subject as well. This work concentrates on creating arbitrary capacitor ratios on a unit-sized common centroid array. To obtain a good layout, a valid mismatch estimation model is necessary. The assumption is that the mismatch is dominated by oxide gradients. From the mismatch variation with the gradient angle or the worst-case estimate, the unit capacitor cells in a rectangular array are assigned to individual capacitors making up the capacitance ratio. The desired ratios are classified into three classes and the assignment is done accordingly; even ratios, odd ratios, and non-integer ratios. With the successful completion of the assignment, the layout is created. Finally, the mismatch from the resulting configuration is calculated.

Another module that is of much interest in analog layouts is "stacked transistors." Stacked transistors are a set of transistors with the drain node of one transistor merged with the source node of the next one. Also, a single transistor can be partitioned into a stack and interleaved with another one. This way, big reductions in area and parasitic capacitance are possible. A methodology for optimum stacked transistor layout generation was first presented in [48] and [49]. The stack generation follows a four-step procedure whereby the first step is identifying and grouping transistors with the same bulk bias node. Then, module splitting is performed on all transistors according to topological adjacency relations and matching constraints. The third step is simply subgrouping transistors with the same widths. The final step is to partition these subgroups into chains.

Partitioning is known to be an NP-complete problem. However, by using analog constraints, a branch-and-bound type approach can be utilized, thus reducing the computation time required. This approach has been developed further in [50], where matching and symmetry constraints have also been enforced. Again, merging devices with different widths has not been considered, but is left to the placement phase of the layout generator. The quality of the stack arrangement is evaluated by a cost function that takes into account critical parasitics and area. Matching constraints are also taken into account when creating the stacks. A separate cost function is created for matching and this cost function drops to zero when all matching constraints are satisfied. Although providing optimal results in terms of the cost function defined, this approach suffers from several drawbacks. One of them is that the human designer can sometimes modify the size of one transistor for better stacking. Another shortcoming of this method is the inability to separate a single larger transistor into several different stacks. An algorithm that relies on predesigned layouts was proposed in [51] to overcome these problems. Another criticism to the optimal stack generator in [50] is that the stacks generated are actually not optimal in terms of convenient routing and balanced parasitics. One approach to alleviating these problems is presented in [52]. This approach uses a hierarchical fragmentation of the circuit to construct the layout. The designer also has the opportunity to select the partition where the optimal stack generation should be done.

The performance optimization constraints include issues related to the port structure in addition to the conventional parameters, thus allowing easier routing at later stages of the layout generation. It is apparent that this subject is far from being resolved and new methods will appear based on more realistic constraints and on more efficient optimization. However, one issue that should always be kept in mind is the closeness of the constraints to reality; that is, how well the parasitics and mismatch effects are being modeled. As the transistor structure gets more complex, modeling these effects becomes more difficult. As pointed out in [53], the performance of the stacked layout can hardly be optimized or even guaranteed if the models are not correct. In this work, expressions for parasitic capacitance of current mirrors, cascode structures, and differential pairs are provided. Furthermore, a more detailed analysis of mismatch in transistors is performed. It is evident that a simple expression for mismatch based only on transistor areas and distance is not enough to model interdigitated, stacked, or common centroid structures.

The transistor stack generators above are very aggressive generators trying to optimize performance parameters and create the best stack partitions for a given circuit. Another type of application for a module generator can be as a design assistant for a full-custom designer. The tool will not be concerned with finding the best partitioning or minimizing multiple costs. The aim of the module generator will be just to save time for the full-custom designer. However, even this is a big problem in itself considering the variety of feasible structures and devices. ALAS! [54]-[56] is such a tool. ALAS! can create layouts for common centroid interdigitated MOS transistors and matched capacitor and resistor pairs based on geometrical specifications like aspect ratio or gate width provided by the user. Even the generation of single transistors can be improved by automating the layout design as demonstrated in the tool FLAG [57]. The transistors created through this software do not have to be rectangular. Although these transistors are not amenable to common centroid and interdigitated structures, they allow for merging at the sides and provide great reductions in area at the cost of worse matching performance and higher parasitic capacitances.

We have chosen to simplify device generation in ALGv2 for several reasons. The main reason is that it is very difficult to extract numerical matching information and/or parasitics for complex layouts. Thus, optimization based on such constraints is somewhat superfluous in our opinion. On the other hand, merging, folding, or interdigitizing devices is of extreme importance in analog design. Thus, we have designed ALGv2 to act on these commands. These commands can be given by an experienced designer who is aware of the trade-offs involved in the layout. Another alternative is the automatic generation of such commands through a sensitivity analysis by the layout advisor software. We are also in the process of developing ALGv3, which will be more performance oriented and will use the sensitivity matrix defined above for device generation. Module layout generator of ALG is responsible for generating all layout structures of the individual devices in the circuit. These devices can be single independent devices (MOS, resistors, capacitors etc.), hybrid devices that are generated by putting two or more devices into a single layout structure (e.g. merged, interdigitized) or independently designed subcircuit modules with ready CIF layout descriptions. We are interested in this section in the generation of single devices or predefined device groups.

A simple transistor generated with ALG is given in Figure 4.1. The current rating for the transistor can be increased. This results in wider drain and source areas for increased current-passing capability, as shown in Figure 4.2.

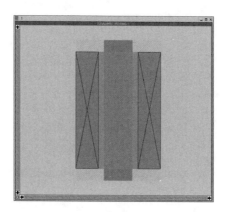

FIGURE 4.1
A simple MOS structure.

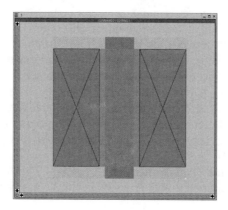

FIGURE 4.2
A simple MOS structure with current rating 2.

Furthermore, a maximum aspect ratio can be defined for the device. Any device which has an aspect ratio greater than this value will be folded as in Figure 4.3.

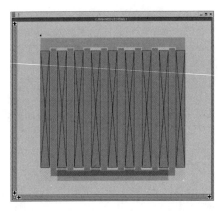

FIGURE 4.3
Folded MOS structure.

Another feature of ALGv2 is its ability to create simple stacked structures. One way of achieving this is to use the "merge" command. If two MOS devices share a common source/drain node, they can be merged together and a single device is produced out of two independent and different devices.

When two MOS devices are merged, the terminals of the independent devices are translated to form the terminal descriptions of the merged hybrid device. The resulting device will consist of a black box and its terminal definitions.

In Figure 4.4, two MOS devices merged as a single device can be observed. It is also possible to generate interdigitated layouts for transistors that have to be matched with each other. An interdigitized MOS structure can be seen in Figure 4.5.

Finally, all of these commands can be combined with each other to form structures where different width transistors can be merged and interdigitized with each other.

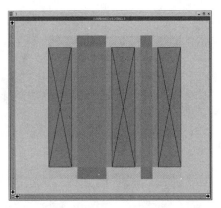

FIGURE 4.4
Two MOS transistors merged as a single device.

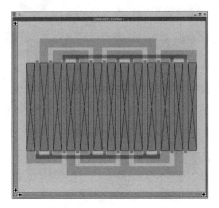

FIGURE 4.5
Interdigitization of two MOS devices.

An additional feature of ALGv2 is to perform self-connections of the devices at the device layout generation stage. Should there be a self-connection on the device, it is performed either by the layout generator or by the grid router. If the connection is straightforward, the device generator performs it. The router is invoked for more complex self-connections of hybrid devices.

Such different layout structures can be seen in Figure 4.6.

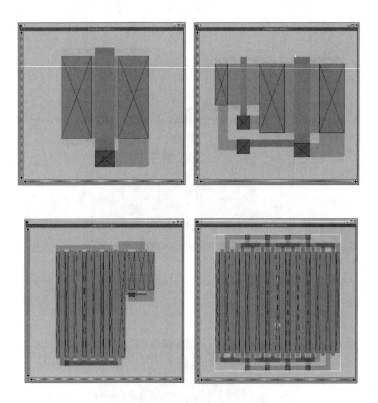

FIGURE 4.6
More complex MOS structures generated by ALG.

It is also possible to read an externally generated CIF layout description and use it as a subcircuit in our design. In such a case, ALGv2 will only read the CIF file to generate the layout that will be used in later steps. No partitioning, placement and routing will be performed for the imported subcircuit. This feature is specially developed for mixed signal design support where ALG will design the analog parts and get the digital parts from a cell library.

After module layout generator is finished, the following are available:

- All individual device layouts, whatever device type it is, are ready.

- Circuit connectivity is calculated.

- Device terminals are generated.

After that point in the design cycle of ALGv2, all devices are assumed to be black boxes whose dimensions and terminals are exactly known.

4.3 Partitioning and Floorplanning

Partitioning plays a key role in the design automation of VLSI circuits, both in the analog and digital domains. It is by no means restricted to layout generation and finds applications in many fields from FPGA's to circuit simulation. Whatever the application is, partitioning a circuit is simply grouping the components or modules of the circuit such that a certain cost is minimized. In general, this cost is simply the number and/or value of the connections across the partitions. The most commonly used subclass of partitioning is when there are only two groups. This type of partitioning is called two-way min-cut partitioning.

Partitioning algorithms can be constructive or iterative. An excellent review of partitioning algorithms for VLSI applications can be found in [58]. It is not our aim to discuss these well-known algorithms here. What will be done in the following paragraphs is to demonstrate how they can be applied to partitioning analog layouts. The simplest bipartitioning (or two-way partitioning) algorithm is the Kernighan-Lin algorithm [59]. Many versions and improvements of this algorithm have appeared in the literature in the following years [60–64]. All of these are based on forming an initial bisection and exchanging nodes across this bisection while reducing the cost. Furthermore, simulated annealing and similar approximate algorithms can also be used for exchanging nodes to minimize the risk of getting stuck at local minima. Constructive algorithms are in general clustering-based. These two types of algorithms can be combined with each other with the constructive algorithm providing a good initial point for further iterative improvement. Although two-way partitioning can seem like a limitation at first sight, recursive application of two-way partitioning is also possible for partitioning into smaller groups [65–67]. The partitioning problem is NP complete and these approximate algorithms may provide some advantage over each other in terms of the quality of solution obtained or the computation time. However, according to our belief, the real success in partitioning of analog circuits lies in the data structure that represents the circuit, the cost function to be minimized, and how the partitions are interpreted in later sections of the layout generator.

ALGv2 uses several different partitioning algorithms. An example to iterative partitioning is the well-known Kernighan-Lin algorithm mentioned above. A constructive partitioning example is the self-organizing neural maps of Kohonen.

The Kernighan Lin (KL) algorithm is a bisectioning algorithm. It starts by initially partitioning the circuit graph into two subsets of equal sizes. Vertex pairs are exchanged across the bisection if the exchange improves the cut size. This procedure is carried out iteratively until no further improvement can be achieved. At each step, KL chooses a pair of vertices whose exchange results in the largest decrease in the cut size or results in the smallest increase, if no decrease is possible. This is done until every vertex in the circuit is tried.

The time complexity of KL algorithm is $O(n^3)$. The KL algorithm is, however, quite robust. It can accommodate additional constraints, such as a group of vertices requiring to be in a specified partition. On the other hand, there are also several disadvantages of the KL algorithm. For example, the algorithm is not applicable for hypergraphs, it cannot handle arbitrarily weighted graphs, and the partition sizes have to be specified before partitioning. Finally, the complexity of the algorithm is considered too high even for moderate size problems. To overcome the disadvantages of KL, several algorithms have been developed. Fiduccia-Mattheyses algorithm and Goldberg-Burstein algorithm are such examples.

Instead of using Kernighan-Lin bisectioning algorithm, ALG can also use Kohonen's Self-Organizing Neural Maps-based multisectioning algorithm. In this case, the circuit connectivity information is converted into vectors to be represented to the Self-Organizing Neural Maps and the locations of the firing neurons are investigated based on the Euclidian distances between each firing neuron to find which device belongs to which subsection. After such a processing, based on the locations of firing neurons, a number of subsections may occur, not only two, as in KL.

In ALGv2, KL and Kohonen SOM partitioning algorithms are used hierarchically. Initially, all the devices are put into a single partitioning node. Selected partitioning algorithm is run for every partitioning node until there is only one device left in each node. In this manner, a partitioning hierarchy tree is constructed.

Let us try to visualize hierarchical partitioning in ALG by using the following example shown in Figure 4.7. Assume that there are five devices present in a circuit. Initially, every device is put into a single partitioning node and KL is applied to that node. At the end of partitioning, devices 2, 3, and 5 are grouped together and devices 1 and 4 are grouped together. Two new partitioning nodes are created for these two groups formed and corresponding devices are put in them. KL is then applied to the node containing 2, 3 and 5. This procedure continues until there are five nodes and every node contains only one device.

Transforming from the tree representation to a floorplan is also quite difficult. At this point there are several alternatives. Even if all blocks are assumed to have a rigid rectangular shape, the decision on whether these blocks could be packed onto a fixed rectangular plane was proven to be NP-

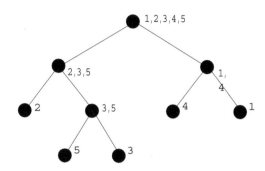

FIGURE 4.7
Hierarchical Kernighan-Lin bisectioning implementation.

complete, while the problem of finding the minimum area packing is NP-hard [68]. The approaches used for this solution have ranged from branch-and-bound to simulated annealing. This issue will be explored in more detail in the following section about placement. With the above constraints in mind, a binary tree translates itself very easily into a slicing floorplan where the modules are organized in a set of slices that recursively bisect the layout horizontally and vertically. It is believed that a slicing representation generally limits the solution space and the optimum solution is, in many cases, not a slicing structure. Recently, it has been demonstrated that binary trees can also be applied to non-slicing floorplans [69]. It has also been proven that the slicing floorplan approach has a time complexity of $O(n \ logn)$ to find the minimum area slicing solution [70]. On the other hand, Lai and Wong [71] argue that, with a suitable combination of slicing trees and post-placement compaction, maximally compact placements can be obtained. Thus, it seems to be a natural selection to use slicing floorplans for analog layout. ALGv2 uses a slicing floorplan and some heuristics to reach the final floorplan.

After the hierarchical partitioning tree is constructed, an estimated area is sliced into smaller portions and each device is given a portion. In this manner, hierarchical partitioning-based initial place assignment is performed as illustrated in Figure 4.8.

A very important point about recursive partitioning is the interpretation of the results. Creating a simple tree directly from the partition is, in general, not enough. Two nodes that are not far from each other in the cost space may fall very far away from each other in the layout just because they were separated from each other in the first partition. Thus, the closeness of the nodes across partitions must be taken into account while forming the tree for the layout. ALGv2 has this feature in forming the tree.

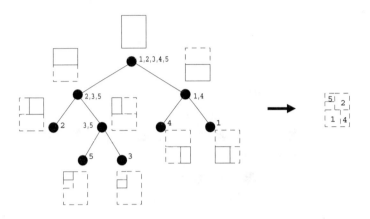

FIGURE 4.8
Floorplanning by means of recursive slicing.

After the initial places are assigned, an optimization on the partitioning tree is performed. Edges that are connected to the same upper hierarchy vertex are swapped, initial area is sliced again, new initial places are assigned and a cost is calculated based on the connectivity and distance criteria. By swapping edges, minimum cost is found and it is accepted as the final partitioning tree. Finalized initial placement is recalculated accordingly.

Figure 4.9 shows an example. Both trees are the same in hierarchy. Different edge configurations result in different initial placements, as shown in the figure. Partitioning tree optimizer in ALG tries to find the best initial placement while not overriding the hierarchy.

In the following paragraphs, an example for partitioning by ALG will be given by using the following opamp circuit in Figure 4.10.

Corresponding to that circuit, ALG will read an input circuit file in the format shown below. This file is in pure SPICE netlist format and contains extra commands for ALG to function parametrically.

```
**
** SPICE file created for circuit bts49-1u
** Technology: scmos
**
**/ PROCESS_LAMBDA 1.0
**/ PROCESS_WELL_TYPE pwell
**/ PROCESS_NAME scmos
*
* Differential input devices
*
```

```
**/ INTERDIGITIZE_DEVICES M1 M2
*
* Current mirror devices at the input
*
**/ INTERDIGITIZE_DEVICES M3 M4
*
* Biasing diode connected devices at the output
*
**/ MERGE_DEVICES M11 M12
*
* Differential input nodes
*
**/ MAIN_CIRCUIT_TERMINAL_NODE 9
**/ MAIN_CIRCUIT_TERMINAL_NODE 10
*
* Output node
*
**/ MAIN_CIRCUIT_TERMINAL_NODE 6
*
* External capacitor connections
*
**/ MAIN_CIRCUIT_TERMINAL_NODE 4
**/ MAIN_CIRCUIT_TERMINAL_NODE 11

**/ CURRENT_RATING default 1
**/ CURRENT_RATING M8 2

M1    7   9   2   2 PFET L=2u W=90u
M2    5  10   2   2 PFET L=2u W=90u
M3    7   7  13  13 NFET L=5u W=82u
M4    5   7  13  13 NFET L=5u W=82u
M5    2   3   1   1 PFET L=6u W=118u
M6    4   5  13  13 NFET L=4u W=191u
M7    4   3   1   1 PFET L=2u W=51u
M8    3   3   1   1 PFET L=6u W=16u
M9   11   6   5  13 NFET L=2u W=40u
M10   6   3   1   1 PFET L=2u W=31u
M11   6   6  12  13 NFET L=5u W=31u
M12  12  12  13  13 NFET L=2u W=25u

.END
```

At the end of the partitioning and initial placement phase, we have the following hierarchy tree generated by ALG. To explain that format more clearly, the left-most numbers indicate the level of hierarchy and the right-most coor-

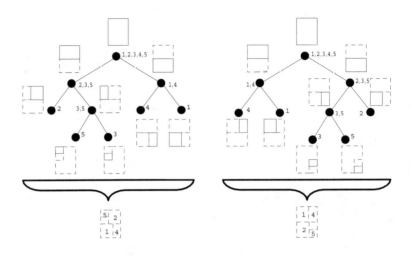

FIGURE 4.9
Hierarchy tree optimization.

dinates indicate the box assigned to that group during the initial placement
phase.

```
Current circuit: bts49-1u
01:-H1 H2 M6 H0 M9 M8 M5 M7 M10 Bounding Box (0,0)-(603,-603)
02:--H1 H2 M6 H0 M9 Bounding Box (0,0)-(603,-407)
03:---H1 H2 M6 Bounding Box (0,0)-(456,-407)
04:----H1 Bounding Box (0,0)-(140,-407)
04:----H2 M6 Bounding Box (140,0)-(456,-407)
05:-----H2 Bounding Box (140,0)-(456,-244)
05:-----M6 Bounding Box (140,-244)-(456,-407)
03:---H0 M9 Bounding Box (456,0)-(603,-407)
04:----H0 Bounding Box (456,0)-(603,-230)
04:----M9 Bounding Box (456,-230)-(603,-407)
02:--M8 M5 M7 M10 Bounding Box (0,-407)-(603,-603)
03:---M8 M5 Bounding Box (0,-407)-(338,-603)
04:----M8 Bounding Box (0,-407)-(107,-603)
04:----M5 Bounding Box (107,-407)-(338,-603)
03:---M7 M10 Bounding Box (338,-407)-(603,-603)
04:----M7 Bounding Box (338,-407)-(492,-603)
04:----M10 Bounding Box (492,-407)-(603,-603)
Ok.
```

To make this textual information more understandable for the reader, Figure 4.11, which depicts the same information, has been included. In this tree

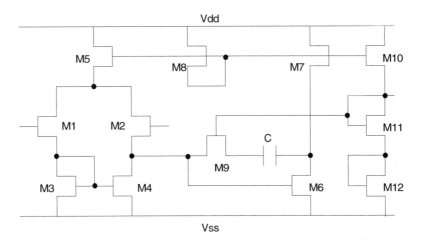

FIGURE 4.10
BTS49 opamp circuit diagram.

representation, the devices indicated by "M" are the MOS, devices whereas the devices indicated by "H" are the hybrid devices produced as a result of interdigitization and/or merging process. In our example, M1 and M2 are interdigitized to produce H0, M3 and M4 are interdigitized to produce H1 and finally M11 and M12 are merged to produce H2. In Figure 4.12, we can see the mapping of the partitioning hierarchy onto the circuit diagram.

At the end of the partitioning phase, an estimated rectangular area is sliced into smaller rectangles and every device is assigned an initial place as explained before. In Figure 4.13, the result can be seen at the layout level.

Other options for floorplanning that do not use trees and slicing could be constructive floorplanning [72] and local search [73]. In [72] a very interesting issue is also brought up. One of the objectives for high-quality layout must be yield enhancement. The authors take this consideration into account while performing the floorplanning.

In analog design, floorplanning is very closely linked with placement. It is sometimes very difficult to determine where one starts and the other ends. More emphasis will be given to placement in this chapter as placement covers floorplanning in many cases. In the next section, different placement algorithms will be discussed and ALGv2 results for placement will be demonstrated.

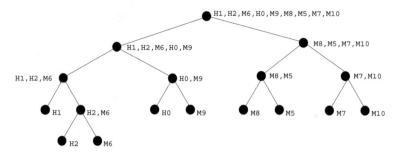

FIGURE 4.11
Hierarchical partitioning tree of BTS49 circuit.

4.4   Placement

Placement of blocks on a chip has been much studied in the literature. The reader is referred to [58] for a discussion of various algorithms. The most notable among these algorithms are branch-and-bound type, simulated annealing-based, and force-directed algorithms. Both branch-and-bound type and simulated annealing-based algorithms try to minimize some cost function. In force-directed placement, the blocks are tied to each other by imaginary springs having constants related to how close they should be to each other. When the system comes to a rest at a minimum energy point, the cost is assumed to be minimized. It should also be noted that human beings perform surprisingly well in packing modules with fixed shapes; however, they are not very good in choosing a shape out of a set of given shapes to achieve an optimal packing. Thus, there is a great need for placement tools to exploit the design space offered by the use of module generators. The above algorithms should be modified to include this idea as well.

Irrespective of the algorithm used, definition of the cost function is very critical in our opinion. The cost function obviously has to involve some kind of area optimization. However, this should not be the only goal, as in many of the floorplanning algorithms presented above. Routing and performance constraints should be added as well if possible. TINA [74] is an example of a tool that generates slicing placements optimal in both area and overall net length. As an example of the importance of including net length in the cost function, TINA reduces the overall net length to one fifth of a net length unaware algorithm, while increasing the area by less than 2% and requiring only twice the CPU time. The algorithm operates by enumerating all slicing configurations that are compatible with the given neighborhood relations. For every possible configuration, the resulting shape and net length are presented to the user,

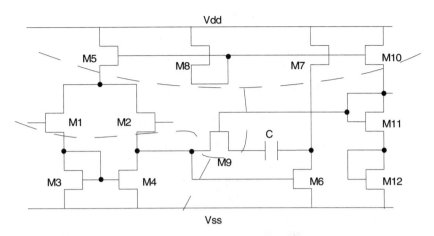

FIGURE 4.12
Mapping of partitioning onto the circuit schematics.

who then selects a configuration. For the selected configuration, the positions and orientations for each module are enumerated again. Obviously, this is far from being a fully automatic tool.

The sequence-pair representation is another representation for the placement problem that is more amenable to simulated annealing type solutions [68]. If sequence-pair representation is used, an initial floorplan is not required. Furthermore, symmetry and device matching constraints can also be added to this representation to increase the quality of the final layout [75], [76]. Simulated annealing can, of course, be applied to a direct absolute representation of the placement, as in KOAN/ANAGRAM II. However, many problems arise in such a representation, especially when the cost function becomes more complex. In a way, it can be said that the classical absolute representation trades off a large number of moves for easier and quicker-to-build layout configurations, whereas topological representations trade off more complex layout constructions for a smaller number of moves [77]. In ALGv2, we have applied a hybrid approach between a slicing representation and a topological representation, as will be described later.

One of the first performance-driven analog placement algorithms was presented in [78]. The authors compare their method to the method of [79] mentioned in earlier sections and to be discussed again in the later sections. In [79], they try to translate performance specifications into parasitic constraints and use these to drive the layout tools. The tool in [78] eliminates the intermediate constraint generation step. The layout tools are directly driven by the performance specifications. The placement tool uses a simulated annealing-based optimization starting from a random point, and takes

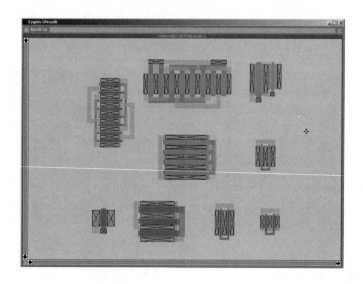

FIGURE 4.13
Initial place assignment of BTS49 circuit at the layout level.

into account symmetry constraints, interconnect parasitics, and device mismatches and combines these with geometrical optimization techniques like merging and abutment. The moves allowed in simulated annealing are device translation, device reorientation, device swap, and device reshape. The cost function is composed of the maximum area, the aspect ratio, overlap, and performance degradation. The performance degradation part of the cost is estimated from the interconnect capacitances and resistances (which are estimated before the routing from a minimum spanning tree) and mismatches. Very simple and approximate formulae are used for all these expressions. Furthermore, routability of the final layout is only indirectly addressed in this cost function.

Zeng et al. [80] advocate the refinement of the placement according to constraints. An initial placement is obtained via geometric optimization. Then, placement refinement is performed based on the constraints. The refinement not only changes positions of modules, but also reshapes and regenerates them if necessary according to the performance constraints. This way, they are able to achieve around 20% area reductions while keeping within the performance constraints.

Another major issue in analog placement is substrate coupling. In standard CMOS technology, high-frequency signals tend to couple across the substrate to other parts of the circuit. Thus, measures for preventing this coupling should be taken and sensitive devices should be placed either in separate wells or far from each other. The detection of substrate coupling and achieving substrate-aware placement can be obtained either by using a simplified circuit model of the substrate [81] or by performing a lengthy electromagnetic simulation of the layout [82], [83]. An issue related to substrate coupling is the insertion of guard rings. Guard rings not only minimize substrate coupling, but also prevent latch-up. The authors in [84] advocate placement of guard rings as a separate step on the generated layout based on certain rules.

ALGv2 performs an initial placement from the floorplan obtained as explained in the previous section. Obtaining the placement from this floorplan is straightforward. Then, a simulated annealing placer is run on this initial placement to refine the placement further.

During SA placement, a cost function is tried to be minimized by introducing disturbances into the system. The cost function is composed of the design rule violations, total silicon area utilization and the total estimated wiring length for the router. The disturbances introduced include small and large translations, rotations and aspect ratio changes of the devices. The cooling profile is parametrical and is currently an exponentially decaying function.

The main role of the simulated annealing placer in ALG is to improve the initial placement. Figures 4.14 and 4.15 below contain two different final place assignments for the BTS49 circuit that we use as an example.

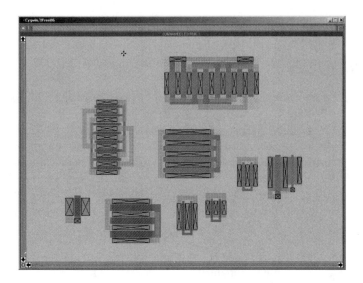

FIGURE 4.14
Final place assignment of BTS49 circuit at the layout level.

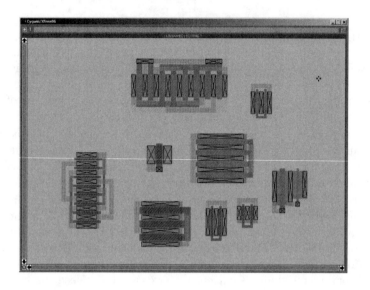

FIGURE 4.15
Alternative final place assignment of BTS49 circuit.

4.5   Routing

The problem of routing arbitrarily placed blocks has been addressed very early in the literature, as it applies to PCB design as well. However, the problem is NP-hard, rendering an optimal solution impossible. Humans have traditionally outperformed computer routers by far. The standard type of router for arbitrarily placed blocks is the maze router, which is guaranteed to find the shortest route between two solutions if it exists. However, the problem begins when more than two nets have to be connected. The optimum path through these is the Steiner tree, finding which is an NP-complete problem. Furthermore, if there is more than one connection to be made, the solution depends primarily on the order in which the connections are made. Finding the optimal ordering and hence the optimal routing becomes a very difficult problem. On top of these difficulties, one should consider the fact that the performance of an analog circuit is heavily dependent on the layout. Thus, the definition of an optimum routing becomes even more difficult.

One of the first routers taking into account these effects was the router in SALIM. The authors claim that a simple cost function consisting of some combination of parasitic resistances, parasitic capacitances, and wire length is not realistic [85]. Furthermore, they state that a sequential approach where wire ordering is correctly addressed may be very efficient. The router that they present performs the routing in two successive steps; symbolic routing and detailed routing. The symbolic router tries to match the electrical constraints, whereas the detailed router is a postprocessing step taking into account technological constraints. In order to perform symbolic routing and electrical optimization, the nets are classified into five functional classes, high sensitivity, low sensitivity, noisy, power supply, and bias. The connections scheduling is achieved by an expert system using information from the floorplanning step.

Another early analog router, which uses a simple cost function and a scheduler, is presented in [86]. The authors utilize a simple maze router for net connections, but the router has a cost function composed of the weighted distance between the nodes, the resistance and capacitance of the connections, a proximity parameter denoting the proximity of the wire to already existing wires, and a congestion parameter of the area. Backtracing of connections is allowed and the connections are scheduled according to several criteria, among which symmetry, user defined priorities, and cost function considerations are the most important. This router is improved in [87] where a constraint generator is used to explore the sensitivities of the critical parasitics. Using this sensitivity information, weights are generated for the area router. The area router in this work, ROAD, is a somewhat more intelligent version of the standard maze router. In ROAD, an efficient scheduler sorts the nets to be routed based on a priority list. The priorities for the individual nets are calculated from the weights for that net. The nets with the tightest electrical requirements are routed first. Further improvements to this approach and a more complete description of the router can be found in [88]. The authors also explain how to partition nets with different current densities and how to use symmetry information more efficiently in the router. Furthermore, they discuss the introduction of shields into the layout. Calculating performance sensitivities will be discussed in a later section about performance-oriented layout generation. However, the interested reader is referred to [89] for a more detailed explanation on using performance sensitivities in routing.

The constraint-based approach outlined above can also be applied to channel routing, especially for mixed signal circuits as explained in [90] and [91]. Channel routing is an easier problem than maze routing and most of the performance information can be incorporated into constraint graphs, which are in turn used to generate the routing. However, we will not pursue channel routing further in this book.

Another approach to analog routing may be to generate a set of candidate routes for each net [92]. Then, the candidate routes can be simultaneously considered for compatibility and a set of compatible routes can be chosen. Furthermore, the nets can be classified according to their types as explained earlier to enhance performance-oriented operation.

One other important issue in routing of analog circuits is current. This issue has been partially considered in some of the aforementioned articles. However, a much more rigorous treatment of current limitations and electro-migration avoidance can be found in [93] and [94]. The main problem in this case is to be able to route lines of different widths successfully. One earlier approach has been to assume uniform minimum width for all wires and correct the thickness later via a postprocessing step. Another approach has been to use a worst-case method in choosing wiring thickness and reduce sizes later. However, it is obvious that both approaches will give suboptimal results.

ALGv2 applies a maze routing algorithm sequentially. Very few routing constraints are defined. However, as explained also in [95], simple maze routing will not work properly even if performance is not an issue. The reason is that maze routing is an algorithm that can solve only one connection properly. If more than one net has to be connected sequentially, many problems appear. We have introduced many heuristics in ALGv2 to overcome these problems.

The maze router in ALG utilizes a two-way propagation. Starting both from the target layer and from the source layer, constrained routing propagation takes place. According to the result of this two-way propagation, the router decides whether to directly connect the source and the target, or to perform single or multiple layer changes.

If the propagation from source meets the propagation from the target at the same layer, no layer change is necessary. If the propagation from source does not meet the propagation from the target at the same layer, the router checks to find out whether those two layers can be connected to each other by placing the appropriate single cut. If so, a suitable location for that kind of cut is searched and it is placed. If this is not the case, or a suitable location cannot be found, a multiple layer change is tried. To perform this task, necessary layer changes are performed and necessary cuts are placed to suitable locations.

ALG's router also utilizes rip-up techniques. If a connection cannot be performed, it randomly rips-up some connections and retries. Selecting which connection to rip-up is currently heuristically performed but a neighborhood-based selection algorithm will be employed.

In the following figures, two instances of the same circuit design are illustrated. The circuit is our original example circuit, BTS49. The designs in Figure 4.16 and Figure 4.17 are the results of the two different runs of ALG for the same circuit. Note the small difference in the final placement and the huge difference in the routing. This is due to the fact that a lot of heuristic decisions are employed in the router of ALG.

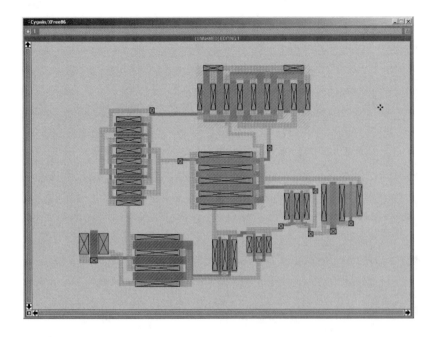

FIGURE 4.16
Final result of ALG for the example circuit.

4.6 Post Layout Improvements

As the careful reader will have observed from the above discussion, most layout generation algorithms will apply a clean-up step after the layout is created. This step will vary from algorithm to algorithm, but will involve some type of DRC (design rule check), some performance optimization, and compaction. Compaction is a well-known problem in IC layout generation, whether analog or digital. A concise review of well-known compaction algorithms can be found in [58]. These algorithms mostly involve generating a constraint graph from which a "rift" is created and eliminating this rift from the layout, thus compacting it. However, the major limitation of compaction algorithms is that they are very successful in one dimension. A two-dimensional compaction is not necessarily the same as applying two one-dimensional compactions successively. Another issue about compaction algorithms is that, although they will reduce the total area, they may end up destroying the

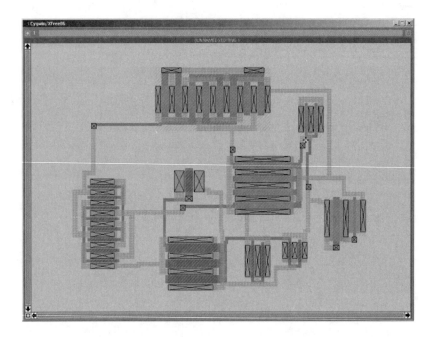

FIGURE 4.17
Another result of ALG for the same example circuit.

symmetry that the layout generator created so painstakingly. Thus, simple compaction algorithms must first be modified to preserve the symmetry associated with the circuit. Okuda et. al. [96] add symmetry constraints into the constraint graph they use for compaction. This way, they are able to make the circuit smaller and at the same time keep it symmetric. However, their algorithm works only in one dimension, namely, the horizontal direction.

In [97], the authors propose the use of LP (Linear Programming) to solve the constraint graph for compaction. The graph is created not only from topological relations as in the standard algorithm, but also from symmetry constraints as in [96]. Furthermore, the compaction algorithm preserves a minimum distance between wires that can cause unwanted signal coupling as well preserving shields. During the compaction, the aim is to make the overall area as small as possible. However, this may not necessarily translate to minimization of every interconnect length or even the minimization of the total interconnect length. These lengths can be very critical in terms of the parasitics they generate. Thus, compaction may turn out to have an adverse effect on the overall performance of the circuit. In [98], the authors advocate the use of a second optimization after compaction to minimize the total interconnection length.

Another variation of the compaction problem is when the blocks of the layout have many shape possibilities. If the blocks are left flexible until this point (which is highly unlikely due to the placement and routing steps preceding compaction) or the blocks are allowed to change shapes and dimensions during compaction interactively with a module generator, compaction becomes much more difficult, but at the same time, much more efficient. One algorithm for the solution of this problem is presented in [99]. It is our belief that this problem, where compaction and shape optimization are carried out together, is not practical because the compaction phase is too late for changing block shapes. Thus, we will not pursue this very interesting problem further in this book.

One other application of compaction can be found in "automatic layout recycling." This term is used to denote the process of porting the layout to a new technology when the dimensions shrink. This problem is much easier for digital circuits, whereas the performance of the analog circuit is highly dependent on many layout factors, one of these being symmetry. Bourai and Shi [100] propose a method for automatic symmetry detection from the original layout. Once this is done, the dimensions of the devices can be shrunk and the layout can be compacted with symmetry constraints in mind as described earlier in this section.

4.7 Performance Issues in Analog Layout Generation

In the previous sections, performance-oriented layout generation was frequently mentioned. It was also explained how various approaches utilize some performance constraints to obtain good layouts. However, the issue of generating performance constraints was not discussed. In this section, we will review several constraint generation methodologies and also explain how performance constraints are generated in our layout advisor (LAD).

Automatic generation of constraints for analog layout was discussed first in the context of analog routing [101]. The constraints in terms of routing boil down to parasitic capacitance and resistance. The authors define the constraint generation problem as generating a set of parasitic constraints from a set of performance constraints. The performance constraints are calculated by using a definition of sensitivity as

$$Sij = [\partial W_i / \partial p_j]_{p_j=0} \qquad (4.1)$$

where, W_i is a performance function whose sensitivity at its nominal value is calculated with respect to a parasitic p_j. For AC sensitivity, the computation requires first the calculation of the DC operating point. However, the computation of AC sensitivity is further complicated by the fact that the result may be a complex number. If this is the case, phase and magnitude sensitivities can be treated differently. Calculating all performance sensitivities with respect to all possible parasitics will identify critical parasitics. These in turn will be used to create the constraints on the layout. The careful reader may note at this point that no information is given about the sign of the sensitivities. In fact, some may be positive and some negative. The human designer can use this fact to reduce or even cancel out some effects. However, the system presented in [101] is not capable of this and utilizes only a worst-case approach. The authors also mention process variations in terms of parasitics only and use this information in matching of parasitics such as capacitance. Also, if matched node-pair information is available, the system in [101] can extract matching constraints for the parasitics. However, creating these sensitivities or constraints will not immediately translate to the routing problem at hand. What has to be done is to create some bounding constraints for the router from the generated constraints and sensitivities. In general, there will be many possible sets of bounding constraints that will satisfy the performance criteria. Selecting the set that allows the maximum flexibility in routing is another issue. To this end, a "routing flexibility function" is defined in [101]. An optimization algorithm is run on this function to create all constraints. The interested reader is referred to [102], where this approach is explained more rigorously by the same authors. The main problem with this approach is that it concentrates only on parasitics brought about by routing.

A different approach is employed in [103] and [104]. Again, sensitivities are calculated, but in a slightly different manner, as below:

$$\sum_i \left(\frac{\delta p_i}{W_u - W_n} \times \frac{\partial(W)}{\partial(p_i)} \right) < 1 \qquad (4.2)$$

where, W is a function in n parameters $p_1, \ldots, p_n$, W_u and W_n are the upper bound and the nominal value of the function, respectively. This sensitivity is called the incremental constrained normalized sensitivity (ICNS). In order to include the effects of process variations on matching devices, a parametric tracking (PT) term is also introduced. PT requirements are also automatically detected in this approach, thus bringing an improvement over the approach of [101] and [102]. The PT and ICNS parameters are used to build a weighted graph from which the layout is deduced.

The constraint generation approach of [101] and [102] can also be generalized for device matching and symmetry considerations [105]. A new set of constraints named topological constraints can be extracted from these considerations and this new set can be added to the already existing constraints.

A major drawback of all the above approaches is the large number of parameters that one has to work with. The problem is not in the generation of sensitivities, although this might require a large amount of computations. There are many algorithms to make this step faster. Furthermore, with ever-increasing computational power, this issue ceases to be a problem. The main issue is how to translate these sensitivities into some format that the layout generator can benefit from. If this is not done correctly, the layout constraints will be either too harsh, thereby making layout generation impossible, or too relaxed, making the final layout incompatible with the performance constraints. Arsintescu and Otten [106] propose a principal component analysis on the sensitivity matrix. This way, sensitivity values having little or no significance on final layout are discarded and only important and independent parameters are retained, thus reducing the problem and making the search for the optimum layout easier.

In our opinion, the first step involved in the process of automatic layout constraint generation is sensitivity analysis. LAD utilizes a sensitivity analyzer, YASA (yet another sensitivity analyzer), which computes the sensitivities of certain performance measures of a given circuit with respect to a set of undesired parasitics that are likely to be introduced in the layout phase.

Our sensitivity analyzer (YASA) accomplishes the task of sensitivity analysis using the perturbation method [102]. In this method, degradation of performance values from nominal and, therefore, their associated sensitivities to layout parasitics, are evaluated by perturbing circuit parameters and performing circuit simulation. YASA incorporates a fast DC simulator that uses the same calculation engine as SPICE. One major advantage that YASA has in doing DC sensitivity analysis is that the integrated DC simulator does not have to be reinitialized for each DC simulation. Instead, the state of the circuit is preserved after each simulation to provide the initial conditions for the next analysis step. Thus, each circuit simulation requires fewer iterations to converge, since all the sensitivity analysis takes place in the vicinity of a certain DC operating point. Hence, DC sensitivity analysis is performed in a very efficient way, yielding results faster.

The sensitivities of AC performance measures are computed in a slightly different manner. Instead of using a generic AC simulator, a simpler calculation engine that makes use of modified nodal analysis (MNA) is developed. Since our analog design automation system is based on a library of certain circuit topologies, it is possible to produce a different MNA "template" for each circuit topology in the library. Using these templates, the calculation engine can generate the modified nodal equations for a certain DC operating point, which are necessary to calculate the values of AC performance measures of concern. The set functions provided by this engine consist of the calculation of complex AC response at a certain frequency and the frequency at which a certain AC response is realized. Using these two functions, YASA can compute the values of many performance measures including gain and phase at a given frequency (spanning the DC range), bandwidth, phase margin and sensitivities of these measures to layout parasitics. The creation of these templates is very simple and can be done automatically. This approach provides faster evaluation as opposed to a generic AC simulator.

YASA takes a standard netlist file as input and, in return, produces a matrix consisting of the sensitivities of a set of performance measures with respect to a set of layout parasitics. The functional structures that make up the two sets of performance measures and parasitics mentioned above are the real actors of YASA's sensitivity analysis. There is a different performance measure structure implementation for each kind of performance measure that can be defined. All performance measure implementations share the functionality of performance value and sensitivity value calculation. What differs among different implementations is the way that the structure is created and configured and how the mentioned values are calculated. Similarly, there is a different parasitic structure implementation for each kind of parasitic that can be defined. These implementations share the functionality of affecting a given circuit in the way they are defined. What is specific to each implementation is the way the target area or element of the target circuit that the parasitic will affect is defined.

When run unassisted by providing only the circuit topology, YASA generates the widest sets of performance measures and parasitics that can be obtained from the provided netlist and performs a thorough analysis. However, sets of performance measures and parasitics can also be flexibly defined by the user to obtain different analysis configurations. There is a set of instructions that make it possible to define these two sets. After performance measures and parasitics are generated, the problem of sensitivity analysis is confined to the following:

```
FOR EACH Performance
{
Performance.ComputeNominalValue
FOR EACH Parasitic
{
  Parasitic.InfectCircuit
  Performance.ComputeParasiticValue
  Sensitivity(Performance,Parasitic)=Performance.Compute-
                                      Sensitivity(Parasitic)
  Parasitic.RestoreCircuit
}
}
```

Currently, parasitic structures that have been implemented and are available in YASA consist of the geometrical parameters of MOSFETs, all MOSFET model parameters and matching parameters that can be derived from these. This means that YASA can compute the sensitivities of performance measures with respect to the parasitics resulting from circuit parameters that are likely to be modified during the layout phase.

Using the sensitivity matrix obtained from YASA, qualitative commands can be given to ALGv2. In a way, the program gives "advice" to the layout generator, thus the name LAD. We have chosen this type of an approach for easy interfacing to the user in that the user can observe and even change the advice given to ALG, thus resulting in different layout structures.

In future versions of ALG (ALGv3 is in development), quantitative advice will be included if the user so chooses. In that case, the sensitivity matrix created by YASA will be processed by ALGv3 to create a layout that conforms to all sensitivity specifications.

4.8 Conclusions and Future Directions

The field of automatic layout generation for analog integrated circuits is by no means a finished and mature field. Quite on the contrary, there are many issues and future directions to be explored that have not been included in the discussion above. One of these issues is reliability. Reliability should be considered at each stage of the layout generation from module generation to routing and final clean-up phases. In [107], reliability issues pertaining to module generation are discussed. Among these are RC time constants for digital circuits, electromigration, current through contacts, voltage drop in interconnections, and parasitic capacitances. The careful reader will have noticed that most of these issues have been considered in the algorithms presented so far, albeit implicitly. A good layout-generation tool should include such reliability calculations more explicitly. An issue closely related and expanding about this concept is fault prediction and layout design for minimum fault probability and maximum testability. In their recent work, the authors of [108] employ the inductive fault analysis technique for fault prediction, whereby physical faults were applied randomly to a large number of different transistor layouts and models based on W, L. These models were further utilized in generating layouts for more complex circuits.

Having discussed at great length many different approaches to analog layout generation, the question of which generates the best layout remains open. Actually, the question gets even more complicated because there is no method measuring how "good" a layout is. Driven by this need, the authors in [109] propose such a quantitative evaluation scheme. Their measure is based on the weighted sum of several factors. The first factor is "area efficiency"; namely, how much of the total area is actually covered by transistors. The second factor is routing optimality, which is a weighted sum of the ratio of the optimal net length to the actual length where the weights are determined by the criticality of the net. The authors have then used this measure for optimization of their layouts. Although a very good idea at first sight, such a measure is inadequate for representing analog circuits since performance issues have not been considered at all.

In [110] and [111], the authors propose a dramatically different approach to circuit synthesis. They note that the synthesis of a circuit followed by layout generation cannot produce optimal results. Layout parasitics should be estimated and compensated for during synthesis. They then explain how this can be carried out. In their synthesis tool, the layout-generation tool is iteratively called and parasitics are estimated. The layout tool calculates the parasitics based on the geometry information and transistor currents. This information is then sent to the synthesis tool along with estimates of routing parasitics and well sizes. Then, the synthesis tool can compensate for the parasitics by sizing the transistors accordingly. Once the calculated parasitics remain unchanged, the final layout is created. The typical iteration count is quite low, typically comprising just a few iterations.

In [112], the same authors discuss the application of their system to analog design for reuse with a case study on a low voltage $\Sigma - \Delta$ modulator. As explained earlier in this book, this is one of the main goals of analog design automation. Their methodology is similar to ours and consists of four major steps. The first step is high-level synthesis where the most suitable architecture and oversampling ratio are chosen followed by modulator coefficient determination. Then, based on this fixed architecture, block parameters are determined. The next step is the synthesis of these blocks. The fourth and last step is the generation of the complete layout following a template-based approach.

References

[1] Itoh, M. and Mori, M., ALE: A layout generating and editing system for analog LSIs, *Proc. ISCAS*, 843, 1990.

[2] Mehranfar, S. W., STAT: A Schematic to artwork translator for custom analog cells, *Proc. CICC*, 30.2.1, 1990.

[3] Conway, J. D. and Schrooten, G. G., An automatic layout generator for analog circuits, *Proc. 3rd European Conference on Design Automation*, 513, 1992.

[4] Walczowski, L. T. et. al., Rapid layout synthesis for analog VLSI, *Proc. ICECS*, 378 1996.

[5] Jingnan, X. et. al., IC design automation from circuit level optimization to retargetable layout, *Proc. ICECS*, 95, 2001.

[6] Berkcan, E. et. al., From analog design description to layout: a new approach to analog silicon compilation, *Proc. CICC*, 4.4.1, 1989.

[7] Berkcan, E. and Currin, B., Module compilation for analog and mixed analog-digital circuits, *Proc. ISCAS*, 831, 1990.

[8] Berkcan, E., MxSICO: A silicon compiler for mixed analog digital circuits, *Proc. ICCD*, 36, 1990.

[9] Borutzky, W. et. al., A novel approach to CAD of analog cells, *Proc. ISCAS*, 1791, 1988.

[10] Bowman, R.J., Analog macrocell layout generation, *Proc. ASIC*, 10-2.1, 1989.

[11] Onodera, H., Kanbara, H., and Tamaru, K., Operational-amplifier compilation with performance optimization, *IEEE JSSC*, 466, 25, 1990.

[12] Harvey, J. P., Elmasry, M. I., and Leung, B., STAIC: A synthesis tool for CMOS and BiCMOS analog integrated circuits, *Proc. ISCAS*, 2004, 1991.

[13] Harvey, J. P., Elmasry, M. I., and Leung, B., STAIC: An interactive framework for synthesizing CMOS and BiCMOS analog circuits, *IEEE Transactions on CAD*, 1402, 11, 1992.

[14] Koh, H. Y., Squin, C. H., and Gray, P. R., Automatic layout generation for CMOS operational amplifiers, *Proc. ICCAD*, 30, 1988.

[15] Shin, H. and Sangiovanni-Vincentelli, A., MIGHTY: A rip-up and reroute detailed router, *Proc. ICCD*, 10, 1986.

[16] Bensoiah, D. A., Mack, R. J., and Massara, R. E., Design assistant approach to analogue layout, *IEE Proceedings – Circuits, Devices, and Systems*, 213, 143, 1996.

[17] Bensoiah, D. A., Mack, R. J., and Massara, R. E., Analog layout generation: from design assistant to automated layout, *Proc. MWSCAS*, 1038, 1997.

[18] Wu, P. B., Mack, R. J., and Massara, R. E., A parameterized block-level layout generation system for CMOS analog ICs, *Proc. ISCAS*, 197, 2000.

[19] Kayal, M. et. al., An Interactive layout generation tool for CMOS analog ICs, *Proc. ISCAS*, 2431, 1988.

[20] Kayal, M. et. al., SALIM: A layout generation tool for analog ICs, it Proc. CICC, 7.5.1, 1988.

[21] Rijmenants, J. et. al., ILAC: an automated layout tool for analog CMOS circuits, *Proc. CICC*, 7.6.1, 1988.

[22] Rijmenants, J. et. al., ILAC: an automated layout tool for analog CMOS circuits, *IEEE JSSC*, 417, 24, 1988.

[23] Wolf, M., Kleine, U., and Schafer, F., A novel design assistant for analog circuits, *Proc. ASP-DAC*, 495, 1998.

[24] zu Bexten, V. M. et. al., ALSYN: Flexible rule-based layout synthesis for analog ICs, *Proc. CICC*, 11.6.1, 1992.

[25] zu Bexten, V. M. et. al., ALSYN: Flexible rule-based layout synthesis for analog ICs, *IEEE JSSC*, 261, 28, 1993.

[26] Garrod, D. J., Rutenbar, R. A., and Carley L. R., Automatic layout of custom analog cells in ANAGRAM, *Proc. ICCAD*, 544, 1988.

[27] Mogaki, M. et. al., LADIES: an automatic layout system for analog LSI's, *Proc. ICCAD*, 450, 1989.

[28] Cohn, J. M. et. al., New algorithms for placement and routing of custom analog cells in ACACIA, *Proc. CICC*, 27.6.1, 1990.

[29] Cohn, J. M. et. al., KOAN/ANAGRAM II: New tools for device-level analog placement and routing, *IEEE JSSC*, 330, 26, 1991.

[30] Kim, Y., Cho, H., and Yoon, K., INALSYS: A layout automation system based on analog layout constraints, *Proc. MWSCAS*, 1209, 1998.

[31] Lee, J.-C., Gowda, S. M., and Sheu, B. J., Fully automated layout generators for high-performance analog VLSI modules, *Proc. TENCON*, 893, 1989.

[32] Hong, S. K. and Allen, P. E., Performance driven analog layout compiler, *Proc. ISCAS*, 835, 1990.

[33] Hong, S. K. and Allen, P. E., Analog circuit layout with optimized performance, *Proc. MWSCAS*, 567, 1990.

[34] Pillan, M. and Sciuto, D., Constraint generation and placement for automatic layout design of analog integrated circuits, *Proc. ISCAS*, 355, 1994.

[35] Malavasi, E. et. al., Automation of IC layout with analog constraints, *IEEE Transactions on CAD*, 923, 15, 1996.

[36] Chen, D. J. and Sheu, B. J., Automatic custom layout of analog ICs using constraint-based module generation, *Proc. CICC*, 5.5.1, 1991.

[37] Chen, D. J. and Sheu, B. J., Generalised approach to automatic custom layout of analogue ICs, *IEE Proceedings – Circuits, Devices, and Systems*, 481, 139, 1992.

[38] Charbon, E. et. al., Imposing tight specifications on analog ICs through simultaneous placement and module optimization, *Proc. CICC*, 24.1.1, 1994.

[39] Gohar, N., Cheung, P. Y. K., and Pun, C. K., RACHANA: an integrated placement and routing approach to CMOS analog cells, *Proc. ISCAS*, 2981, 1992.

[40] Cohn, J. M. et. al., Techniques for simultaneous placement and routing of custom analog cells in KOAN/ANAGRAM II, *Proc. ICCAD*, 394, 1991.

[41] Prieto, J. A. et. al., An algorithm for the place-and-route problem in the layout of analog circuits, *Proc. ISCAS*, 491, 1994.

[42] Prieto, J. A. et. al., A performance-driven placement algorithm with simultaneous Place and Route; optimization for analog ICs, *Proc. IEEE Design Automation and Test in Europe Conf. (DATE)*, 389, 1997.

[43] Kubo, Y. et. al., Explicit expression and simultaneous optimization of placement and routing for analog IC layouts, *Proc. ASP-DAC*, 467, 2002.

[44] Lin, Z. M., Huang, Y-J., and Hsiau, K-H., LAKE: a performance-driven analog CMOS cell layout generator, *Proc. IEEE Asia-Pacific Conference on Circuits and Systems*, 564, 1994.

[45] Yufera, A., Rueda, A., and Huertas, J. L., Flexible capacitor and switch generators for automatic synthesis of data converters, *Proc. ISCAS*, 3162, 1991.

[46] Yufera, A. et. al., ALSAC, an automatic layout system for successive approximation converters, *China International Conference on Circuits and Systems*, 407, 1991.

[47] Sayed, D. and Dessouky, M., Automatic generation of common-centroid capacitor arrays with arbitrary capacitor ratio, *Proc. IEEE Design Automation and Test in Europe Conf. (DATE)*, 576, 2002.

[48] Malavasi, E., Pandini, D., and Liberali, V., Optimum stacked layout for analog CMOS ICs, *Proc. CICC*, 17.1.1, 1993.

[49] Liberali, V., Malavasi, E., and Pandini, D., Automatic generation of transistor stacks for CMOS analog layout, *Proc. ISCAS*, 2098, 1993.

[50] Malavasi, E. and Pandini, D., Optimum CMOS stack generation with analog constraints, *IEEE Transactions on CAD*, 107, 14, 1995.

[51] Spanoche, S-A. and Arsintescu, G. B., Stack based module generator for analog MOS circuits, *Proc. International Semiconductor Conference*, 139, 1996.

[52] Naiknaware, R. and Fiez, T. S., Automated hierarchical CMOS analog circuit stack generation with intramodule connectivity and matching considerations, *IEEE JSSC*, 304, 34, 1999.

[53] Zeng, X. et. al., Parasitic and mismatch modeling for optimal stack generation, *Proc. ISCAS*, 193, 2000.

[54] Bruce, J., Li, H., and Dallabetta, M., ALAS!: an analog layout assistant for matched and balanced CMOS components, *Proc. ASIC*, 267, 1995.

[55] Bruce, J. D. et. al., Analog layout using ALAS!, *IEEE JSSC*, 271, 31, 1996.

[56] Pease, R. A., Comments on Analog layout using ALAS!, *IEEE JSSC*, 1364, 31, 1996.

[57] Mathias, H. et. al., FLAG: A flexible layout generator for analog MOS transistors, *IEEE JSSC*, 896, 33, 1998.

[58] Sherwani, N. A., *Algorithms for VLSI Physical Design Automation*, 3rd Ed., Boston, MA: Kluwer 1999.

[59] Kernighan, B. W. and Lin, S., An efficient heuristic procedure for partitioning graphs, *Bell System Tech. Journal*, 291, 49, 1970.

[60] Fiduccia, C. M. and Mattheyses, R. M., A linear-time heuristic for improving network partitions, *Proc. DAC*, 175, 1982.

[61] Krishnamurthy, B., An improved min-cut algorithm for partitioning VLSI networks, *IEEE Transactions on Computers*, 438, C-33, 1984.

[62] Dutt, S., New faster Kernighan-Lin type graph partitioning algorithms, *Proc. ICCAD*, 370, 1993.

[63] Saab, Y. G., A fast and robust network bisection algorithm, *IEEE Transactions on Computers*, 903, 44, 1995.

[64] Cheng, S-W and Cheng, K-H, ENISLE: an intuitive heuristic nearly optimal solution for mincut and ratio mincut partitioning, *Proc. ISCAS*, 167, 2001.

[65] Wei, Y. C. and Cheng, C. K., Toward efficient hierarchical designs by ratio-cut partitioning, *Proc. ICCAD*, 298, 1989.

[66] Hagen, L. and Kahng, A., Fast spectral methods for ratio-cut partitioning and clustering, *Proc. ICCAD*, 10, 1991.

[67] Cong J. and Lim, S. K., Multiway partitioning with pairwise movement, *Proc. ICCAD*, 512, 1998.

[68] Murata, H. et. al., VLSI module placement based on rectangle-packing by the sequence pair, *IEEE Transactions on CAD*, 1518, 15, 1996.

[69] Balasa, F., Modeling non-slicing floorplans with binary trees, *Proc. ICCAD*, 13, 2000.

[70] Shi, W., An optimal algorithm for area minimization of slicing floorplans, *Proc. ICCAD*, 480, 1995.

[71] Lai, M. and Wong, D. F., Slicing tree is a complete floorplan representation, *Proc. IEEE Design Automation and Test in Europe Conf. (DATE)*, 228, 2001.

[72] Prasad, R. and Kohen, I., Constructive floorplanning with a yield objective, *Proc. International Symposium on Quality Electronic Design*, 261, 2001.

[73] Adya, S. N. and Markov, I. L., Fixed-outline floorplanning through better local search, *Proc. ICCD*, 328, 2001.

[74] Abthoff, T. H. and Johannes, F. M., TINA: analog placement using enumerative techniques capable of optimizing both area and net length, *Proc. EURO-DAC*, 398, 1996.

[75] Balasa, F. and Lampaert, K., Module placement for analog layout using the sequence-pair representation, *Proc. DAC*, 274, 1999.

[76] Balasa, F. and Lampaert, K., Symmetry within the sequence-pair representation in the context of placement for analog design, *IEEE Transactions on CAD*, 721, 19, 2000.

[77] Balasa, F., Device-level placement for analog layout: an opportunity for non-slicing topological representations, *Proc. ASP-DAC*, 281, 2001.

[78] Lampaert, K., Gielen, G., and Sansen, W. M., A performance-driven placement tool for analog integrated circuits, *IEEE JSSC*, 773, 30, 1995.

[79] Charbon, E., Malavasi, E., and Sangiovanni-Vincentelli, A., Generalized constraint generation for analog circuit design, *Proc. ICCAD*, 408, 1993.

[80] Zeng, X. et. al., A constraint-based placement refinement method for CMOS analog cell layout, *Proc. ISCAS*, 408, 1999.

[81] Lin, C. and Leenaerts, D. M. W., A new efficient method for substrate-aware device-level placement, *Proc. ASP-DAC*, 533, 2000.

[82] Mitra, S. et. al., Substrate-aware mixed-signal macro-cell placement in WRIGHT, *Proc. CICC*, 24.2.1, 1994.

[83] Mitra, S. et. al., Substrate-aware mixed-signal macrocell placement in WRIGHT, *IEEE JSSC*, 269, 30, 1995.

[84] Ker, M-D. et. al., Automatic methodology for placing the guard rings into chip layout to prevent latchup in CMOS IC's, *Proc. ICECS*, 113, 2001.

[85] Piguet, S. et. al., A new routing method for full custom analog IC's, *Proc. CICC* 27.7.1, 1990.

[86] Malavasi, E., Chilanti, M., and Guerrieri, R., A general router for analog layout, *Proc. CompEuro*, 5-49, 1989.

[87] Malavasi, E., Choudhury, U., and Sangiovanni-Vincentelli, A., A routing methodology for analog integrated circuits, *Proc. ICCAD*, 202, 1990.

[88] Malavasi, E. and Sangiovanni-Vincentelli, A., Area routing for analog layout, *IEEE Transactions on CAD*, 1186, 12, 1993.

[89] Choudhury, U. and Sangiovanni-Vincentelli, A., Use of performance sensitivities in routing of analog circuits, *Proc. ISCAS*, 348, 1990.

[90] Choudhury, U. and Sangiovanni-Vincentelli, A., Constraint-based channel routing for analog and mixed analog/digital circuits, *Proc. ICCAD*, 198, 1990.

[91] Choudhury, U. and Sangiovanni-Vincentelli, A., Constraint-based channel routing for analog and mixed analog/digital circuits, *IEEE Transactions on CAD*, 497, 12, 1993.

[92] Sajid, K. et. al., A global routing methodology for analog and mixed-signal layout, *Proc. ASIC/SOC*, 442, 2001.

[93] Lienig, J., Jerke, G., and Adler T., Electromigration avoidance in analog circuits: two methodologies for current-driven routing, *Proc. ASP-DAC*, 372, 2002.

[94] Lienig, J., Jerke, G., and Adler T., AnalogRouter: a new approach of current-driven routing for analog circuits, *Proc. IEEE Design Automation and Test in Europe Conf. (DATE)*, 819, 2001.

[95] Mani, N. and Quach, N. H., Heuristics in the routing algorithm for circuit layout design, *IEE Proceedings – Computers and Digital Techniques*, 59, 147, 2000.

[96] Okuda, R. et. al., An efficient algorithm for layout compaction problem with symmetry constraints, *Proc. ICCAD*, 148, 1989.

[97] Felt, E. et. al., An efficient methodology for symbolic compaction of analog ICs with multiple symmetry constraints, *Proc. EURO-DAC*, 148, 1992.

[98] Felt, E. et. al., Performance-driven compaction for analog integrated circuits, *Proc. CICC*, 17.3.1, 1993.

[99] Okada, K., Onodera, H., and Tamaru, K., Compaction with shape optimization, *Proc. CICC*, 24.6.1, 1994.

[100] Bourai, Y. and Shi, C-J. R., Symmetry detection for automatic analog-layout recycling, *Proc. ASP-DAC*, 5, 1999.

[101] Choudhury, U. and Sangiovanni-Vincentelli, A., Constraint generation for routing analog circuits, *Proc. DAC*, 561, 1990.

[102] Choudhury, U. and Sangiovanni-Vincentelli, A., Automatic generation of parasitic constraints for performance-constrained physical design of analog circuits, *IEEE Transactions on CAD*, 208, 12, 1993.

[103] Gad-El-Karim, G. and Gyurcsik, R. S., Generation of performance sensitivities for analog cell layout, *Proc. DAC,* 500, 1991.

[104] Gad-El-Karim, G. and Gyurcsik, R. S., Use of performance sensitivities in analog cell layout, *Proc. ISCAS*, 2008, 1991.

[105] Charbon, E., Malavasi, E., and Sangiovanni-Vincentelli, A., Generalized constraint generation for analog circuit design, *Proc. ICCAD*, 408, 1993.

[106] Arsintescu, B. G. and Otten, R. H. J. M., Constraints space management for the layout of analog IC's, *Proc. IEEE Design Automation and Test in Europe Conf. (DATE)*, 971, 1998.

[107] Wolf, M. and Kleine, U., Reliability driven module generation for analog layouts, *Proc. ISCAS*, 412, 1999.

[108] Prieto, J. A. et. al., An approach to realistic fault prediction and layout design for testability in analog circuits, *Proc. IEEE Design Automation and Test in Europe Conf. (DATE)*, 905, 1998.

[109] Wu, P. B., Mack, R. J., and Massara, R. E., A quantitative method for evaluating the quality of a layout, *Proc. MWCAS*, 1140, 2000.

[110] Dessouky, M., Lourat, M-M., and Porte, J., Layout-oriented synthesis of high performance analog circuits, *Proc. IEEE Design Automation and Test in Europe Conf. (DATE)*, 53, 2000.

[111] Dessouky, M. and Lourat, M-M., A layout approach for electrical and physical design integration of high-performance analog circuits, *Proc. ISQED,* 291, 2000.

[112] Dessouky, M. et. al., Analog design for reuse-case study: very low-voltage $\Sigma - \Delta$ modulator, *Proc. IEEE Design Automation and Test in Europe Conf. (DATE)*, 353, 2001.

Chapter 5

Design Automation Case Studies

5.1 Introduction

In this chapter, we illustrate the analog design automation approach presented in this book with three examples. As stated in Chapter 1, these examples were selected from different domains to illustrate the generality and validity of our approach. All designs were completed to chip level and were verified. The authors are aware of the fact that simple presentation of the results obtained will not suffice in illustrating and explaining a design automation system. Hence, care was taken to document various stages of the design and explain various design decisions taken by the human and the design automation system. The reader should also keep in mind that more examples were given in the previous chapters about the working of each tool composing the whole design automation system. Hence, the number of examples presented in this book is not limited to just the three case studies presented here, but is quite a sizable number, when the examples in the previous chapters are considered. One final remark about the examples is that each one illustrates a different design strategy. This fact demonstrates the versatility of the tool set described throughout the book, where designers have the option of using some tools interactively and not using some of them, but making the design themselves.

The first example presented is the switched-capacitor filter design example. This case study was chosen because it represents a member of the class of discrete time analog systems. It is also a well-known and well-studied type of circuit, with extensive literature and design examples. Furthermore, it is a relatively easy circuit to design. Thus, we have opted for a fully automated design flow for this example. First, the high-level synthesis tool was run in conjunction with the performance estimator to obtain the block diagram and the block specifications. Next, the opamps, which are the major blocks, were synthesized through the circuit-level synthesizer. Finally, the layout was generated for the SC filter. The whole design was performed with minimal manual intervention.

The second example presented is the analog neural network design example. This circuit is a representative of the class of analog mathematical blocks. Again, a high-level tool was used to determine the block diagram and specifications for each block. Since an expert designer in this field was available, some blocks were designed by the expert designer at the transistor level while some were designed automatically. All layouts except for the opamp were hand crafted. This particular case study emphasizes various points. One of these is the ability of the analog design automation system to design such mathematical blocks. Another point is that designers can use their expertise if necessary to intervene and perform some designs themselves wherever it is necessary or wherever the designer feels comfortable. Again, the design flow is top-down with manual design being a portion of the overall effort, mostly in the lower stages of synthesis. A final important comment about this design is that it is the only automated analog neural network design in the literature, to the best of our knowledge.

The third and final case study presented in this chapter is the design of A/D converters. There are actually two designs presented under this heading, flash and pipelined converters. In these designs, only the high-level automation tool was utilized. For the flash converter, the tool was used to generate the block diagram, calculate some component values, and most importantly, to select between predesigned standard blocks. For the pipelined converter, the high-level tool was used to explore the design space and evaluate various alternatives. Again, predesigned blocks were used for some blocks and new circuits were designed for some other blocks. These designs and selections were performed interactively with the high-level design automation tool. Once the block level and circuit level designs were completed, the layouts were hand crafted.

It should be noted that the application of the design automation system is not limited to only these three classes of circuits. Adding a new circuit class to the design automation system depends on creating a high-level design automation tool for that class. Furthermore, if designers choose to carry out the high level design by themselves, they can utilize the remaining tools to create designs for new classes of circuits.

In the remainder of this chapter the three case studies outlined above will be discussed. Section 5.2 will describe in detail how the SC filter design is carried out, whereas Section 5.3 will be concerned with the analog neural network design. Finally, the A/D converter design will be discussed in Section 5.4.

5.2   SC Filter Design Example

The first case study in this chapter is related to a particular class of discrete-time analog systems; switched-capacitor filter implementations are pervasive in many signal processing applications and they are also well studied with an accompanying plethora of literature and design examples. Owing to this fact, a fully automated design flow has been adopted in this case study. In particular, a high-level synthesis tool was developed and run in conjunction with the performance estimator first, in order to obtain the block diagram and the block specifications. Next, the opamps, which are the major blocks, were synthesized through the circuit-level synthesizer. Finally, the layout was generated for the SC filter. The whole design was performed with minimal manual intervention.

Figure 5.1 depicts the conceptual flow diagram pertaining to the design automation environment formed for this application. In addition, this dia-

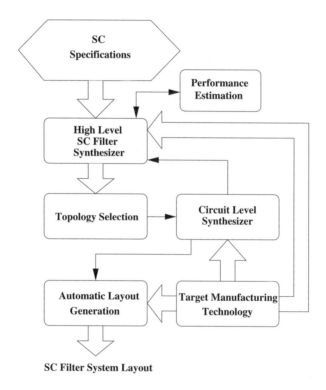

SC Filter System Layout

FIGURE 5.1
Fully automated SC filter design automation flow.

gram is in concurrence with the generic VLSI design automation flow shown in Figure 1.4, although there are slight differences due to the customizations required for the application at hand. It is also worthy to note that, to further optimize the overall design in terms of area and complexity, the synthesized opamp is back-annotated to the high level optimizer. This mechanism facil-

itates the incorporation of realistic nonidealities so that the finalized layout
can be verified without further design iterations.

5.2.1   Design Specifications and High-Level Synthesis

The following specifications are considered for a SC bandpass filter design
application.

$$f_{clock} = 100 \text{ kHz}$$
$$f_{center} = 1600 \text{ Hz}$$
$$Q = 25$$

The target silicon implementation technology was a 1.5μm standard n-well
CMOS technology with double-poly and double-metal capability with 5V operation. The double-poly feature provides poly1/poly2 sandwich capacitor
formation such that a unit capacitor was chosen to be 24μm x 24μm yielding
323 fF for the target implementation technology. Maximum allowable capacitance ratio was then chosen to be 200. As discussed in Chapter 2, this ratio
is very important in determining the minimum and maximum values for the
capacitors.

Upon providing these specifications, the SC filter high-level synthesizer
produced a fourth-order filter (2 biquads cascaded, see Figure 1.5 for a biquad
schematic) formed with BTS opamps. The following capacitance values were
generated.

$$C_a = 11.628 \text{ pF}$$
$$C_b = 61.37 \text{ pF}$$
$$C_c = 646 \text{ fF}$$
$$C_d = 11.951 \text{ pF}$$
$$C_f = 323 \text{ fF}$$
$$C_h = 323 \text{ fF}$$
$$C_{total} = 86.241 \text{ pF}$$

For the BTS opamp shown in Figure 5.2, the block specifications were produced as

$$A_o > 250$$
$$R_{out} < 22k\Omega$$
$$BW > 11kHz \text{(for 1\% gain reduction at } f_{center})$$

Since the bandwidth of the filter is quite small (1600/25=64Hz), gain stays
almost constant in this interval although the bandwidth of the BTS was less
than 11kHz. It was also observed that a gain greater than 250 at the center frequency was sufficient. Next, these specifications were entered to the

circuit-level synthesizer as constraints, which include the open-loop gain, BW, CMRR, phase-margin, output resistance, output offset voltage, and power dissipation:

```
Aogood=10000 Aobad=250 BWgood=50e3 BWbad=11e3
CMRRgood=200000 CMRRbad=10000 PWdisgood=0.1E-3 PWdisbad=10E-3
OffVgood=1E-3 OffVbad=.3 PMgood=100 PMbad=50 RoGOOD=15E3 RoBAD=40e3
```

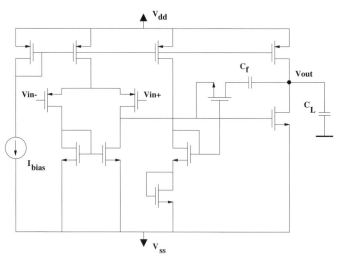

FIGURE 5.2
Schematic of the BTS opamp.

5.2.2 Circuit and Layout Level Synthesis Results

In the resulting circuit, the compensation capacitance was 3pF, which resulted in a decrease in the bandwidth from 11kHz to 4kHz. Specifications of the synthesized BTS with no load capacitor are provided in Table 5.1. The frequency response of the synthesized BTS opamp is shown in Figure 5.3.

Following the design automation flow shown in Figure 5.1, the SC filter was then resynthesized using the specifications of the above opamp and switches formed of CMOS transmission gates with W=9.6μm, L=1.6μm and on-resistance of $R_{on} = 10k\Omega$. The second pass of the high-level synthesizer taking into account the nonidealities of switches and opamps yielded the following updates on the biquad capacitance values.

$$C_a = 10.013 \text{ pF}, C_b = 52.003 \text{ pF}, C_c = 323 \text{ fF}, C_d = 6.137 \text{ pF},$$

$$C_f \ \& \ C_h = 323 \text{ fF}, C_{total} = 69.122 \text{ pF}.$$

Most importantly, this resulted in a 17pF reduction in total capacitance, which was mainly due to the increased BTS opamp gain. Simulation results of the SC filter at the system, circuit and layout levels are given in Figure 5.4. For

TABLE 5.1 Specifications of the Synthesized BTS.

A_o	5457 (74.7dB), 5000 around center frequency
BW	4kHz
PM	85°
R_{out}	23.8kΩ
V_{offset}	5mV
CMRR	30650 (89.7dB)
Power	5.6mW
$C_{feedback}$	3pF
Output Swing	-2.36V / 2.5V
Area	308μm x 207.2μm = 63817.6 μm^2
Slew Rate	90V/μs

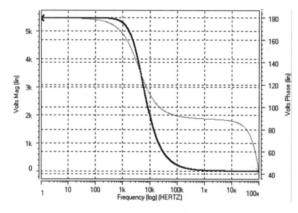

FIGURE 5.3
Frequency response of the synthesized BTS.

each input frequency, a 40 ms transient analysis was performed to ensure that the desired steady-state response is obtained. The resulting time domain waveforms were inspected. The layout of the BTS opamp was then generated automatically. Figure 5.5 displays the layout of the opamp. Circuit extraction was performed on the layout, taking into account parasitics. This extracted circuit was resimulated. The overall performance of the automatically synthesized filter is displayed in Table 5.2 for various levels of abstraction.

TABLE 5.2 SC Synthesis Results.

Level	f_{center}	BW	Q
System	1600	64	25
Circuit	1600	64	25
Layout	1602	65	24.6

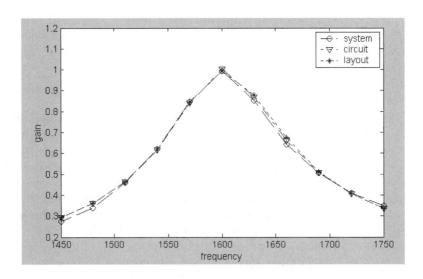

FIGURE 5.4
SC filter system simulation results.

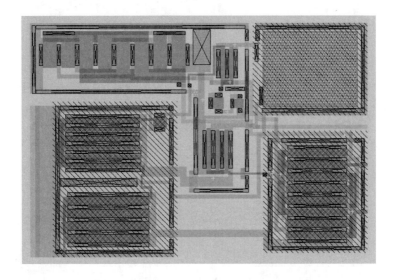

FIGURE 5.5
Layout of the automatically generated opamp.

Inspection of Table 5.2 shows very little degradation with layout parasitics. If so desired, the extracted performances of the opamps and switches can be resubstituted to the system-level synthesis tool, which can calculate slightly modified capacitance values for exact fit. However, taking into account process variations and model inaccuracies, the small error was deemed to be insignificant.

As outlined earlier, the main goal of minimizing the overall design iterations have been attained in this case study by observing that all three levels are in close agreement. Finally, the generated layout of the SC filter has an area of 745.6μm x 1134.4μm = $845809\ \mu m^2$. This layout is included in a test chip for prototype fabrication. The chip layout is provided in Figure 5.6. As a final remark, the software system was run on a PC with 1.47GHz AMD Athlon Processor. The high level SC synthesis tool required 30 seconds, whereas it took 15 minutes for the circuit-level synthesizer to produce the transistor level circuits with an evaluation rate of 1275 circuits/sec. The BSIM3 models of the target CMOS technology are provided in Appendix A.

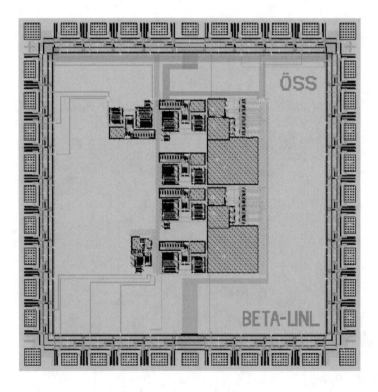

FIGURE 5.6
Test chip layout of the SC bandpass filter.

5.3 Neural Network Design Example

Once the performance criteria for building blocks of ANN are obtained as described in Chapter 2, a decision has to be made in order to select a set of constraints for each building block, namely the synapse, opamp, and the sigmoid. For a given set of available block topologies, circuit-level design of these blocks has to be performed next. For the synapse and the sigmoid, this job has been carried out manually as described in the design flow of Figure 5.7. Initially, a hand analysis is carried out to determine the DC

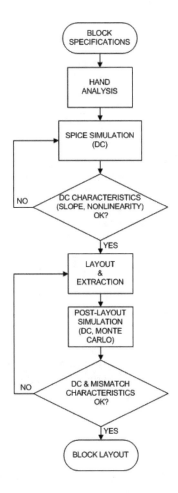

FIGURE 5.7
Circuit-level design flow for the synapse and the sigmoid.

bias points for the given technology parameters, then the DC characteristics of the block are realized in the first loop. The main aim of this step is to achieve the required linearity and operational range. Precaution has been taken for identically sized transistor pairs in differential pairs and current mirrors to keep the mismatch between those transistors at a low level: All transistors have been designed using non-minimum gate length, moreover, transistor widths at output stages have been kept "large." Then, the layout of the block is designed in a full-custom manner where analog layout techniques for mismatch prevention have been applied: large transistors are split into smaller ones; guard rings are put along the boundaries of wells for better well-contacts; transistors in pairs are laid out closely and parallel to each other. After extracting the layout, a few small corrections have been applied to transistor sizing to cancel offsets. Finally, for each block, 100 Monte Carlo runs have been simulated that include device-based mismatch effects in their model parameters. The results of these Monte Carlo runs are checked to determine whether the maximum deviations in DC characteristics are within the range of the specifications as dictated by the synthesis tool. If necessary, the dimension of transistors can be increased without changing the width-over-length ratio so that the mismatch performance is enhanced while the DC characteristics are preserved.

Adopting this design flow, a decision has to be made about the set of constraints as block specifications to be supplied by the system level modeling tool of Chapter 2. For the three example problems considered in Chapter 2, an investigation yields that, for satisfactory performance (100% success for XOR and 3-bit parity problems and 84% success for 2-D classification problem) the block level specifications should be as follows.

$$Synapse\ nonlinearity : 10\%$$
$$Synapse\ mismatch : 5\%$$
$$Opamp\ offset : 10\ mV$$
$$Sigmoid\ mismatch : 5\%$$
$$Expected\ success\ rate : 84\%$$

The deviation of the sigmoid block is not relevant for satisfactory training. However, the maximum value at sigmoid output is also limited by supply voltages hence we can assume that any deviation up to 50% (maximum output of 1.5V) is acceptable.

5.3.1   ANN Building Blocks

Several representative circuits will be described in this section, which can be used as building blocks for analog ANN. These circuits have also been implemented as a prototype chip in a $1.5\mu m$ CMOS technology. The synapses in a neural network can be realized by analog multipliers if the inputs and the weights can be represented by voltages. The synapse circuit used in the chip is a modified version of the well known Gilbert multiplier [1, 2] as shown in Figure 5.8. The inputs are in the form of voltage differences and are denoted by $(x_2 - x_1)$ and $(y_2 - y_1)$. The output of the original Gilbert multiplier is a current difference $(U - T)$ and this difference is converted to a single-ended current (Z) through current mirrors. This improves the linearity of the multiplier as well as providing easy interfacing to the following circuitry. Using voltage input – current output synapses for analog ANN is very suitable for VLSI implementation since the actual signals from outside are mostly in voltage form, and the summing operation on synapse outputs can be performed by connecting the synapse outputs together. As the design flow of Figure 5.7 has been followed, a synapse circuit has been designed that is aimed at satisfying the system level specs. The layout of the designed synapse is given in Figure 5.9 and the dimensions of transistors in the synapse are given in Table 5.3, where W and L represent the width and the length of the transistors, respectively.

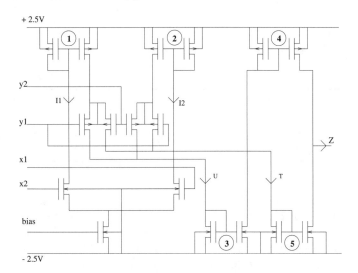

FIGURE 5.8
Circuit diagram of the synapse.

The use of current-output synapses enables the summation of those currents by simply connecting them together at the input of the neuron. This current sum can be converted to voltage by using an opamp similar to the one employed in Section 5.2 and a resistor as a current-to-voltage converter. Another function of the opamp is to fix the voltage of the summation node so that nonlinearities in the addition due to finite output resistances of the

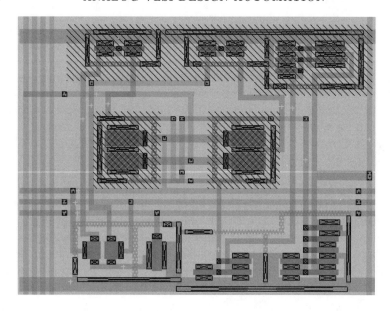

FIGURE 5.9
Layout of the synapse circuit $(460\mu m \times 400\mu m)$**.**

TABLE 5.3 W/L Ratios of Transistors in the Synapse $(\mu m/\mu m)$.

Mirror number					Diff. pair		bias
1	2	3	4	5	x	y	
12.8/3.2	12.8/3.2	12/3.2	38.4/3.2	24/3.2	6.4/12	23.2/11.2	19.2/6.4
12.8/3.2	12.8/3.2	24/3.2	41.6/3.2	60/3.2	6.4/12	23.2/11.2	19.2/6.4

synapses are minimized. The layout of the opamp was generated automatically. This layout is provided in Figure 5.10, whereas the pertaining synthesis specifications of the opamp are given in the next section. The I-V conversion is usually required since most activation function blocks in the literature require voltage input. The essential requirements for opamp based I-V conversion are that the opamp exhibits a large input impedance (to satisfy the virtual ground requirement), and the opamp can supply enough current from its output stage.

A sigmoid generator introduced in [3] is used following the opamp to generate the activation function for the neuron. This generator is depicted in Figure 5.11 where the layout of the sigmoid block can be seen in Figure 5.12.

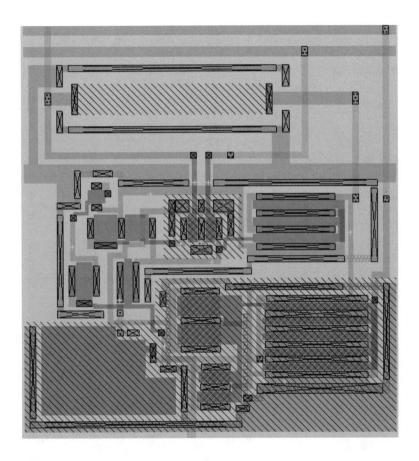

FIGURE 5.10
Layout of the opamp ($400\mu m \times 430\mu m$).

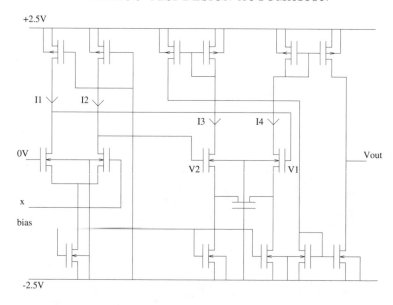

FIGURE 5.11
Sigmoid circuit.

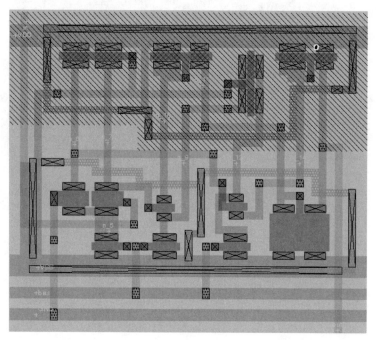

FIGURE 5.12
Sigmoid layout ($330\mu m \times 275\mu m$).

Transistor sizing is determined in the same manner as described for the synapse. Next, the synapse and sigmoid are simulated in Hspice. Characteristics of the synapse and sigmoid blocks are given in Figures 5.13 and 5.14 respectively. The simulation values are displayed as marks and the regression curves to those data points according to the modeling formulae shown in Equations 2.10 and 2.14 are drawn in solid. The regression expressions are repeated for convenience in Equations 5.1 and 5.2.

$$\mu(x, w) = c_0 + c_1 xw + c_2 x^3 w + c_3 xw^3 \tag{5.1}$$

$$\phi(x) = d_0 + \frac{d_1}{1 + exp(d_2 x)} \tag{5.2}$$

Nonlinear regression employing the Levenberg-Marquard method has been applied for fitting the simulation data to the above mentioned equations. The results are summarized in Tables 5.4 and 5.5.

TABLE 5.4 Regression Results for the Synapse.

Parameter	Value	Uncertainty
c_0	-0.749E-06	0.597E-07
c_1	0.647E-04	0.362E-06
c_2	-0.659E-05	0.259E-06
c_3	-0.153E-04	0.223E-06
Maximum absolute error: 6.98E-6		
Proportion of variance explained: 0.9983		

TABLE 5.5 Regression Results for the Sigmoid.

Parameter	Value	Uncertainty
d_0	-1.09	0.418E-02
d_1	0.227	0.714E-02
d_2	-0.3696	0.2559E-01
Maximum absolute error: 0.0247		
Proportion of variance explained: 0.9999		

The high values of variance explained indicate that the data points can be approximated efficiently by the curves. This can be interpreted as a guarantee that the operations carried out by system-level synthesis tools are indeed meaningful. As the models used in the hardware training by software are realistic, physical realizations would also exhibit a close agreement with the training results.

To identify the nonlinearity of the synapse, as defined in Equation 2.11 and displayed graphically in Figure 2.17 in Chapter 2, the slope of the curve for maximum allowed weight ($w_{max} = 1V$) at $x = 0$ is calculated numerically. Then, a straight line is extrapolated through the origin, having this slope of 49.5E-6 A/V. The nonlinearity of the synapse is the ratio of the difference between the "ideal" synapse output (extrapolated value of 49.5E-6 A) and the actual value (43.3E-6 A) at the maximum allowed input ($x_{max} = 1V$), over the ideal value. This comes out to be 13.6% for the synapse characteristics of Figure 5.13. Moreover, the Monte Carlo simulations reveal that the synapse mismatch is around 3.5%

For the sigmoid block, as shown in Figure 5.14, the maximum value where the output saturates is $1.17V$. The deviation of the sigmoid block, as defined in Equation 2.15 and displayed graphically in Figure 2.18 in Chapter 2, is the difference between this maximum value and the maximum ideal value of $1V$. Hence, the deviation is $0.17V$ or 17%.

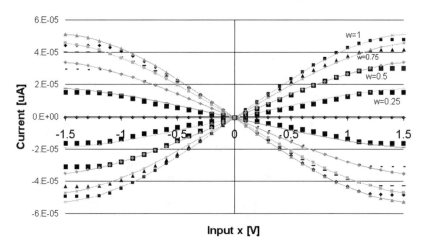

FIGURE 5.13
Synapse characteristics.

Sigmoid Characteristics

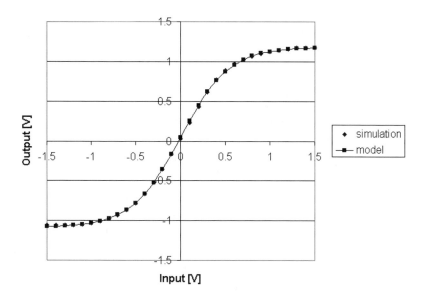

FIGURE 5.14
Sigmoid characteristics.

5.3.2   Synthesis Test Results

Prior to testing various applications, a BTS opamp was first synthesized for the analog ANN. The specifications for the NN BTS are as follows.

$$A_o > 400$$
$$R_{out} < 55k\Omega$$
$$BW > \text{a few hundred Hz}$$
$$V_{offset}: \quad \text{as low as possible.}$$

These specifications were entered to the circuit-level synthesizer as constraints, which include the open-loop gain, BW, CMRR, phase-margin, output resistance, output offset voltage, and power dissipation:

Aogood=10000 Aobad=400 BWgood=1e3 BWbad=.1e3
CMRRgood=200000 CMRRbad=10000 PWdisgood=0.1E-3 PWdisbad=5E-3
OffVgood=1E-3 OffVbad=.3 PMgood=100 PMbad=50 RoGOOD=15E3 RoBAD=40e3

Specifications of the synthesized BTS with no load capacitor are provided
in Table 5.6. Next, the three problems discussed in Chapter 2 have been

TABLE 5.6 Specifications of the
synthesized BTS for ANN Applications.

A_o	6388 (76dB),
BW	306Hz
PM	86°
R_{out}	53.6kΩ
V_{offset}	-3.7mV
CMRR	58153 (95dB)
Power	1.55mW
$C_{feedback}$	1.5pF
Output Swing	-2.44V / 2.5V
Slew Rate	7V/μs

simulated using the electrical models of the ANN blocks to verify the validity
of our approach. To that end, several different sets of weight values obtained
by the system-level synthesis tool are used. As the designed synapse does not
satisfy our specification of nonlinearity, the system-level synthesis tool is run
again to get the estimated success rate for the nonlinearity level of 13.6%.
The results of circuit simulation-based forward runs are given in Table 5.7
where, at each run, weights obtained by the system-level tools are used. The
last line resembles the blocks' characteristics being designed.

For XOR and 3-bit parity problems, on the other hand, all patterns can
be successfully identified for the cases where the system-level synthesis tool
resulted in convergence. Specifically, for the given nonlinearity and mismatch
factors of the building blocks, convergence could be achieved.

The tools were run on a PC with 1 GHz Pentium-III CPU and 256 MBytes
of RAM and a Sun Ultra Sparc 5/270 workstation with 128 MBytes of RAM.
It took around 15 min. for the BTS opamp to be synthesized on the PC.
Synapse and sigmoid circuits required additional preparatory overheads due
to their introduction to the synthesis environment. Consequently, expert-
defined equations were derived in 1 hour per block, whereas SPICE and DC
optimization iterations took approximately 2 hours on the PC. In addition,
layout and extraction required 6 hours per block on the workstation. Finally,
Monte Carlo simulations required 4 hours including netlist preparations and
simulation runs. A test chip was also created that houses an analog ANN
composed of the blocks mentioned in this section. In particular, the test chip

TABLE 5.7 ANN Results.

synapse nonlinearity	mismatch in synapse (%)	opamp offset(mV)	mismatch in sigmoid (%)	expected success rate (%)	simulated success rate (%)
20	10	20	10	48	59
20	10	5	10	79	72
20	10	5	5	83	85
20	5	5	1	87	84
20	1	5	1	85	79
10	10	20	10	46	57
10	10	5	10	83	77
10	5	10	5	84	82
10	5	5	1	91	83
10	1	5	1	93	78
5	10	20	10	57	60
5	10	10	10	65	61
5	10	10	5	81	83
5	5	5	1	94	81
5	1	5	1	92	76
13.6	3.5	3.7	4.1	87	86

has four inputs, three neurons and outputs. The overall design time required to complete the chip was less than 3 days. Figure 5.15 displays the layout of the CMOS chip, which includes this system realization. The BSIM3 models of the target CMOS technology are provided in Appendix A.

5.3.3 Concluding Remarks

The ultimate aim of the work was to be able to design ANN systems in a top-down approach where system level requirements have to be transformed into building blocks with predictable performance characteristics. As is evident from simulations, an important step toward that goal has been taken. The system-level synthesis tool specifies the individual performance requirements of building blocks. Then, using a mixture of custom and automatic design methodologies, ANN building blocks with desired characteristics can be obtained. A simulation-based optimization approach, as described in Chapter 3, is utilized during the synthesis of blocks at circuit level. If the blocks do not satisfy the requirements of the higher-level synthesis tool, there is still the possibility of estimating the performance of the system based on the designed building blocks. A more intelligent optimization can be carried out for selecting the optimal combination of block specifications that can be based on a cost function including terms related to area and power.

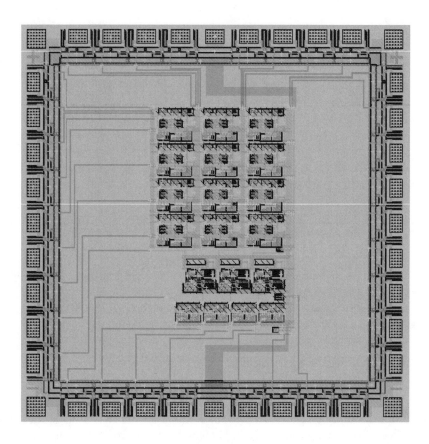

FIGURE 5.15
Test chip layout of the analog ANN.

5.4   A/D Converter Design Example

The third and final case study presented in this chapter is the design of A/D converters. There are actually two designs presented under this heading, flash and pipelined converters. In these designs, only the high-level automation tool was utilized. For the flash converter, the tool was used to generate the block diagram, calculate some component values, and the most important, to select between predesigned standard blocks. For the pipelined converter, the high-level tool was used to explore the design space and evaluate various alternatives. Again, predesigned blocks were used for some blocks and new circuits were designed for some other blocks. These designs and selections were performed interactively with the high-level design automation tool. Once the block-level and circuit-level designs were completed, the layouts were hand crafted.

As opposed to the previous two examples, an automated top-down approach was not employed in the design of A/D converters. There are several reasons for the approach utilized here. The first and foremost reason is that A/D converter design is a very critical design area where every block has to be designed to millivolt accuracy. The tools presented in this book are capable of doing any design, but many aspects of them need to be improved to get critical designs right at the first pass. These aspects were discussed in the respective chapters. The second reason for not utilizing a fully automated approach is simply to illustrate to the reader the power and different applications of the described tools. As mentioned several times earlier, these tools offer the user the flexibility of operating in a fully automated mode or in an interactive mode. Furthermore, the tools are clearly separated from each other in a well-defined and logical manner so that the human designer can replace some of the tools in any step. Thus, the presented tools can work in many different design flow strategies.

In our A/D converter design example, we have opted for an interactive design approach at the system level and a full custom design at the circuit and layout levels. Since the circuit and layout design automation was already demonstrated in the previous two examples, we do not believe that this is a shortcoming, but rather an advantage, as it shows how easily the human designer can integrate into the proposed approach. Furthermore, this allows the use of some predesigned blocks from other designers or from some previous designs of the same designer.

The A/D converter design example actually consists of two examples, a flash and a pipeline converter example. These two designs are discussed next.

5.4.1 Flash Module

The block diagram of a flash converter was presented in Chapter 2 of this book, where some general comments on the operation principles can also be found. In the same chapter, the models used for the high-level synthesis of flash converters were explained in detail. The main set of components in the flash ADC is the comparator bank. It dominates the power and area equations. However, the remaining parts of the ADC are also important for selecting the most efficient design if more than one pre-designed comparator is available or if the comparator will be designed from scratch either by the human or by the ADA tools. As mentioned above, we have chosen not to use a design automation tool for comparators. On the contrary, we have assembled a library of comparators, some of which have been designed by other designers. Then, the high-level design problem boils down to choosing the correct comparator and the optimal resistor values for the specified solution.

Five comparators were available to us and information about their characteristics, determined by simulations, were added to the comparator library used by the design automation tool. Among the circuit characteristics, supply current, area, offset voltage, offset bias current, and rise and fall times can be mentioned, to name a few. Each of these characteristics controls one or more specification for the A/D converter. For example, the rise and fall times of the comparator for a fixed capacitance defined by a single digital standard cell input capacitance give us the slew rate information for the comparator. However, it is very difficult to isolate and model every effect and some were ignored or intentionally left out for future modeling in improved versions of the tool. One such important effect in the comparator is "kick back noise." This effect will probably bring some capacitance restrictions on the resistor chain and some gain restrictions on the preamplifier stage of the comparator. However, careful simulations of the extracted circuit from the final layout did not demonstrate too many problems related to this effect.

The very first design decision is to select the process. The chosen one is an AMS CMOS technology (Austria Mikro Systeme International AG - 0.6 μm Process with 5V, p-substrate, 3-metal, 2-poly, high resistive poly), which allows the use of three metal layers for routing and clock distributions, two poly layers for efficient capacitor designs and a high resistive poly layer that eases the design of resistors in a limited silicon space. To have more headroom for large number of bits, a 5V supply was chosen for the ADC design.

In order to design a test chip, a run has been done for the specifications presented in Table 5.8. The methodology selects 33Ω (poly) or 1200Ω (High resistive poly) sheet resistance and assigns proper W and L values considering the mismatch characteristic.

The tool has different modes of operation. One can be just checking the feasibility of the design. If comparators are available in a library file so that their characteristics are known, the design specifications can be given and the program checks the feasibility of the design. In addition to this feature, the

TABLE 5.8 Flash Converter Specifications.

Specification	Value
Resolution	6 and 7
Reference Voltage ($V_{ref_+} - V_{ref_-}$)	1 to 2
Unit Resistance in the Resistor Ladder	60 to 1200
Clock Frequency	Set to 20MHz
Supply Voltage	Set to 5V
Max Area	Set to minimum
Max power	Set to minimum

tool can fill in the missing design parts. For example, the user can provide the comparator and operating voltage, frequency, and other specs while leaving the unit resistor in the resistor string. This way, the optimum resistance value for this design can be found automatically. However, this is just an example. The user can utilize the tool in many different levels of automation.

For the test chip run, another flow was used. A library of comparators was provided to the tool and the ranges for resolution, resistance, reference voltage, area and frequency were given as shown in the Table 5.8.

First of all, resolution is very important and high-resolution rates are very hard to achieve for flash architectures. For the test design (prototype), a resolution of 6 bits was selected. The chosen technology enables us to use either poly resistance or high resistive poly resistance. These differ in sizes and mismatch characteristics. Also, they have different systematic error variances, which are actually not very important for Flash ADC's. Since the resistances are used as voltage dividers, the more important issue in the design is not their absolute values, but their ratios. However, the systematic error in the resistance values may lead to incorrect power estimations. For AMS technology, systematic errors for high resistive poly are quite large, although it has better mismatch behavior. In test design, the tool gives two different design suggestions, one with ordinary poly resistance and one with high resistive poly resistance with cost values very close to each other. Our tool is not able to model systematic errors fully. Thus, we have chosen to utilize ordinary polysilicon as resistance in the test chip to decrease the effect of the systematic errors whose effects are not modeled perfectly. Thus, the matching behavior was worsened. To alleviate this effect, extra care was taken in drawing the layout of the resistors. The models in the software using the mismatch characteristics provided by the manufacturer show the matching errors to be within limits for the given specifications

The clock frequency was chosen as 20 MHz for this test design. However, this choice is not a final one because the final design outperforms this specification by far. The reason is that the clock frequency is mainly determined by the comparators, and those in the library had very good speed performances. In testing the chip, the frequency will be swept over a large range to obtain the actual operating frequency. Initial simulations show that comparator works satisfactorily up to 50MHz in a flash ADC.

The high-level design automation tool also gives the opportunity to observe the effect of the reference voltage on resistor values as well as power and area consumptions. For the test run, the lower limit was chosen as 1V and the voltage has been swept up to 2V. The tool gives an optimum value of 1.4V for minimum power and area, but this value can be modified for different input signal ranges. The comparator, in our case, should work linearly with the changing input voltage. Simulations showed that the comparators designed for test run are quite linear if the input changes between 2 and 3 volts.

After a successful run, the tool selects a configuration that utilizes "Comparator 1" as a result of a cost function evaluation. The coefficients of the cost function are equally weighted. Some other coefficients may result in smaller areas but higher power consumptions, or vice versa.

The near optimum result that the methodology gives for each comparator is given in Table 5.9. In this table, the first column shows the label of the

TABLE 5.9 Various Comparator Characteristics.

Comparator	Characteristics of the comparator	Number of configurations suggested by the tool
1.	Small current and offset voltage	16
2.	Best rise and fall time	23
3.	Min. offset voltage	46
4.	Min. supply current, high offset voltage	1
5.	Average design	22

comparator. The next column explains the characteristic differences of the comparator compared with the others. The last column shows the number of different solutions with different resistor, power, and area values, that satisfy the given specifications. Since power and area values were not specified, more than one solution can be possible. Then, cost function selects the optimum one.

The tool selects the comparator, which has a 1.41mA supply current, 10mV offset voltage, 3.9ns fall time and 5.4ns rise time. It has a dimension of 186μm x 420μm. The area of the comparator is quite large. The reason is to overcome the mismatch issues that can be very problematic in such ADC designs. Monte Carlo simulations show that the offset of the comparator can vary more than 100% for extreme cases. However, the probability of this situation can be overcome by careful design. Although the layout has been done by using dummy elements and symmetry techniques, the devices are drawn larger than required in order to further increase the tolerance to the process parameters. This seems to be intelligent if this comparator is used many times and in different places of the layout. One good design may overcome many mismatch problems throughout the circuit. The disadvantage of large devices lies in area and operation speed. The former is within acceptable limits while the latter was found to be well within specifications, as mentioned above.

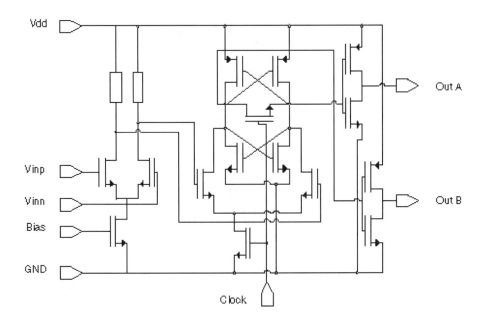

FIGURE 5.16
Schematic of the comparator.

"Comparator 1" shown in Figure 5.16. can be divided into three blocks. The first one is the preamplifier stage that gives the advantage of reducing offset and kick-back noise. This also leads to smaller sizes in the second block. This second block is a clocked latch circuit with positive feedback. The last stage consists of inverters that adjust timing and signal levels. Sizing of the second stage is quite important. Larger sizes give faster responses but the gate capacitance is then quite large. This causes some problems with the preamplifier stage. For our design, we have carefully sized the transistors through manual design. The circuit design was inspired by [4], but has been modified by the authors. The layout of this comparator is displayed in Figure 5.17. Time simulation results verifying the performance are shown in Figure 5.18. For inputs below the reference level, the output essentially stays high while slightly fluctuating between 5V and 4.7V due finite bandwidth and charge-sharing effects. Once the input transition occurs, the clocked output varies between 4.7 and 0V.

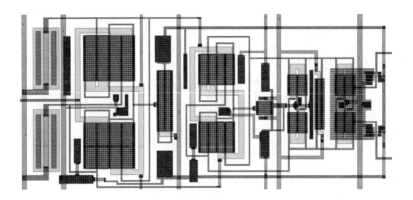

FIGURE 5.17
Layout of the comparator.

Since the chip is for test purposes, some digital logic was added to the encoder part. In this manner, the outputs without encoding are multiplexed. This gives the advantage to inspect the output of each comparator. In case of a process error, the output with error can be found. Although this logic brings some extra power and area for the digital circuitry, these values are quite small and they are tolerable. The output for the area and power is given in Table 5.10.

TABLE 5.10 Breakdown of Area and Power into Modules.

p_{tot}:	Total estimated power (W)	$4.5176e^{-01}$
p_{res}:	Estimated power for resistors (W)	$5.4688e^{-05}$
p_{comp}:	Estimated power for comparators (W)	$4.5120e^{-01}$
p_{dig}:	Estimated power for digital circuit (W)	$5.0100e^{-04}$
a_{tot}:	Total estimated area (mm^2)	$9.6386e^{-06}$
a_{res}:	Estimated area for resistors (mm^2)	$6.2720e^{-11}$
a_{comp}:	Estimated area for comparators (mm^2)	$4.9997e^{-06}$
a_{dig}:	Estimated area for digital circuit (mm^2)	$3.5504e^{-07}$
W:	Width of a unit resistance (m)	$7.0000e^{-07}$
L:	Length of a unit resistance (m)	$1.4000e^{-06}$
sheet:	Unit resistance selected (Ω)	1200
Costs:	Minimum value achieved from cost function evaluation	$1.0904e+02$

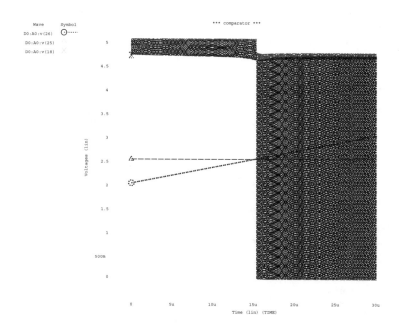

FIGURE 5.18
Clocked comparator with 10mV offset.

MentorGraphics Suite was used for designing the circuit. For the analog part, Accusim, Design Architect, Ic Station, Schematic Generator, and Calibre were used. Also, simulations were crosschecked with Hspice. For the digital part, VHDL was used and simulations were done with Modelsim and QuickSim II. The designed circuits were then synthesized by Leonardo. The analog and digital parts were connected in Ic Station. Area estimations are quite similar with the layout of the test chip that is actually nearly $9.3mm^2$. The estimated value can be modified by the routing constant. This constant increases the area estimation by a ratio about 1.2–1.5, which represents the area consumed by routing. Since the testing is not finished completely, the power estimations have not been checked yet. The power information should be gathered from simulations and total estimated power cannot be checked with simulation values. The layout of the design is shown within the test chip displayed in Figure 5.26.

5.4.2   Pipeline Module

As discussed in Chapter 2, pipeline ADC's are medium or high resolution ADC's. They have relatively slow operation frequency compared with Flash ADC's, but a higher resolution. For the test chip, a resolution of 9 bits is selected as seen in the specifications displayed in Table 5.11, since a high-performance ADC is not required. The aim is to test the model and such resolutions are adequate for our purpose. Higher resolutions consume higher silicon area and more circuitry must be added for trimming or calibration purposes.

TABLE 5.11 Pipeline Converter Specifications.

Specifications	
Resolution	9 bits
Redundant bits	First stage 0, Last stage 1, Others 1 bit
Capacitor value in MDAC	8pF
Resistor value in Sub_ADC	1200 Ω
Total Area	Set to minimum
Total Power	Set to minimum
No. of stages that need calibration	Set to minimum
Max. number of stages	18

After deciding the effective resolution, the tool can now determine the possible configurations for pipeline stages. The number of stages and resolution of each stage is important. Higher-stage resolutions lead to a smaller number of stages but the ADC in each stage grows exponentially, causing a similar growth in the power consumption, area, and design difficulty. This is the first tradeoff to be considered.

The next tradeoff is the operation mode of the pipeline stages; that is, whether they should be single-ended or differential. A differential design methodology eases the design of the pipeline converter by eliminating the effect of component offsets at the end of each stage. However, it increases power and area. We have opted for a fully differential design. Hence, there are two capacitor arrays and amplifier is differential. But there is only one flash converter.

First of all, the tool finds all possible configurations. For the specifications given above for the test design, the configurations are listed below.

```
1 1 1 1 1 1 1 2
2 1 1 1 1 1 2 0
2 2 1 1 1 2 0 0
2 2 2 1 2 0 0 0
3 1 1 1 1 2 0 0
3 2 1 1 2 0 0 0
3 2 2 2 0 0 0 0
3 2 2 2 0 0 0 0
3 3 1 2 0 0 0 0
3 3 3 0 0 0 0 0
4 1 1 1 2 0 0 0
4 2 1 2 0 0 0 0
4 3 2 0 0 0 0 0
4 3 2 0 0 0 0 0
```

Please note that, in this list, the numbers give the number of bits converted in each stage, as explained in Chapter 2. Hence, the first entry in the list corresponds to an eight-stage design with only the last stage being a 2-bit conversion with all others being 1-bit conversions. As another example, the last entry corresponds to a three-stage design with the first stage being a 4-bit converter, the next stage, a 3-bit converter and the final stage only a 2-bit converter.

Redundant bits are generally used in each stage to reduce errors. The careful reader will note that errors that occur in earlier stages will be very important not only because they cause errors in the more significant bits, but also because they propagate to later stages. In our tool, the stages are divided into three categories. These are the first stage, last stage and the others between. For some cases, redundant bits may not be used for the last stages because they mostly do not require correction or the errors are insignificant, as explained above. In our design, we have decided not to use redundant bits in the first stage despite the fact that the errors may be significant. The intermediate stages were chosen to have one redundant bit each and the last stage has none. Furthermore, the digital logic following the pipeline stages was decided to be designed such that the outputs were multiplexed and the results without correction could be both observed to test our model. With these restrictions in mind, the feasible architectures are modified to the following list:

```
1 2 2 2 2 2 2 3
2 2 2 2 2 2 3 0
2 3 2 2 2 3 0 0
2 3 3 2 3 0 0 0
3 2 2 2 2 3 0 0
3 3 2 2 3 0 0 0
3 3 3 3 0 0 0 0
3 4 2 3 0 0 0 0
3 4 4 0 0 0 0 0
4 2 2 2 3 0 0 0
4 3 2 3 0 0 0 0
4 4 3 0 0 0 0 0
4 4 3 0 0 0 0 0
```

Each stage contains an ADC, a DAC, an adding element, and a gain element. However, all these functionalities can be realized by utilizing a single MDAC and a Flash ADC. The MDAC contains a gain element (typically an OTA), a capacitor array, and switching elements. In our test design, MUXs and transmission gates are used in order to realize this switching logic.

The next step was forming a library of amplifiers in the MDAC. In addition to checking the amplifiers available in the library, our tool was used to calculate the limits for the amplifier dictated by the overall specifications. After comparison of the results, it was decided to design a new amplifier for this ADC design. Again, as a design choice, this amplifier was not designed automatically, as illustrated in Chapter 3, but was manually designed.

Another one of the elements in the MDAC is the capacitor array. The tool asks for the total capacitor value for the array. For our design, it was selected as 8pF. This directly translates to area requirements via the technology file containing capacitor value for unit area. Total capacitance value is important for noise calculations and accurate charge transfer. Larger values need higher amplifier gain to satisfy the accuracy requirements.

Since the total capacitance value is known, noise calculations for the MDAC can be done. The test design has a resolution of 9 bits without redundant bits. Resolutions of less than 12 bits do not create input referred thermal noise that is effective in the configuration selection. For the test design, all configurations have tolerable thermal noise values, but some configurations have relatively smaller values. These configurations are those with fewer numbers of stages as smaller stage numbers lead to smaller noise values. However, in that case, stage resolutions increase, causing an exponential increase in area and power consumption. To select the most suitable configuration, the tool utilizes a cost function. The cost function evaluates all its contributors, among which are power, area, noise, and number of stages that require calibration. Each parameter in the cost function is multiplied by a user-defined constant to customize the optimum solution for different requirements. In test design, one of the most important parameters is the number of stages that require calibration. Since no correction logic is supplied in order to inspect output of each stage, this number should be small as possible. Thus, its coefficient is assigned to be larger than the others.

The cost function gives configurations 322223 and 3333 as two possible solutions. However, designing a 3333 configuration is simpler since each stage is exactly the same as the others. Since we have adopted a manual design approach at the circuit and layout levels, we have chosen a 3333 configuration for the test chip.

The tool also gives minimum amplifier gain, minimum accuracy needed by the capacitor array, and also the gain of each MDAC. The first-stage minimum gain was given as 61 dB. The following stages require smaller gain values. From the library of available amplifiers, those with smaller area and power can be selected for the latter stages. However, we have chosen to use the same amplifier all through the test design. Although this will increase the area slightly, it will cause ease and uniformity in the design, especially at the layout level. In this manner, a 3333 configuration will be more advantageous than the others. The designed amplifier is a two-stage amplifier with compensation capacitance. However, the output DC level should still be very accurate even though it was designed for differential use. Thus, common mode feedback was employed for the circuit. Extensive simulations for the amplifier demonstrate a gain of 78dB without any gain-boosting techniques, which is sufficient for the test design. Employing common mode feedback can be a challenging task. In the test design, common mode signal was taken from the output and isolated from the output by source follower. The signal can be too large for a feedback signal. So, a gain stage was added to attenuate the feedback signal. This circuitry supplies the required accuracy needed. The schematic diagram of the amplifier is given in Figure 5.19, whereas its layout is presented in Figure 5.20. Some sample simulation results for the amplifier are depicted in Figures 5.21 and 5.22.

This amplifier was used in a differential MDAC, as explained above. Although a conceptual block diagram of a single-ended MDAC was discussed in Chapter 2, a more complete block diagram of the differential MDAC is given in Figure 5.23 for the sake of clarity for the reader.

Some sample simulation results for the MDAC and its layout can be found in Figures 5.24 and 5.25 respectively. Figure 5.24 shows the gain of 4 for a 20mV differential input.

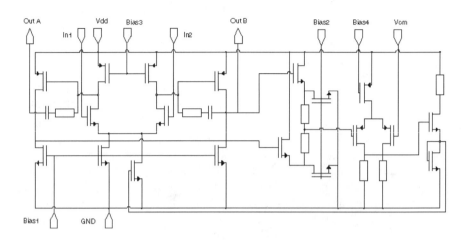

FIGURE 5.19
Schematic diagram of the amplifier.

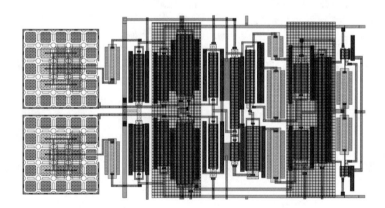

FIGURE 5.20
Layout of the amplifier.

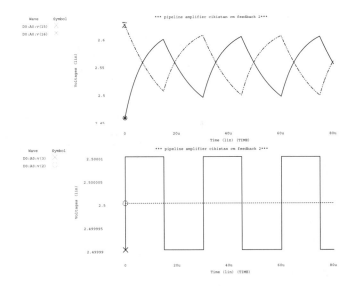

FIGURE 5.21
The output for 10 mV swing showing the charging response time.

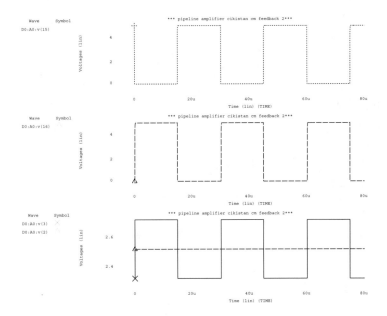

FIGURE 5.22
The output for 2 mV swing showing the charging response time.

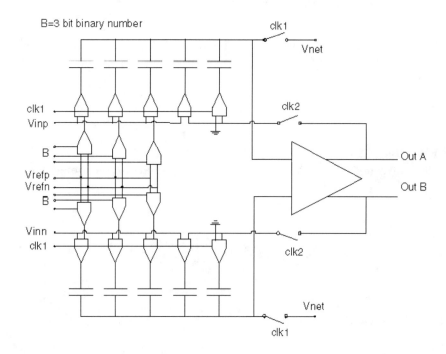

FIGURE 5.23
Block diagram of the differential MDAC.

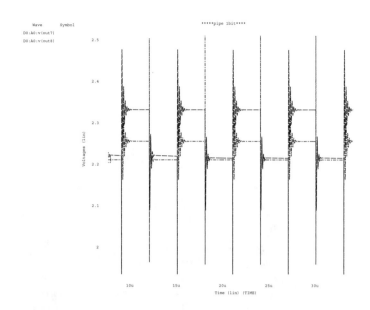

FIGURE 5.24
Gain for a 20 mV differential input.

The layout of the overall pipelined converter occupies $8.9mm^2$ and was designed again via the MentorGraphics suite of tools. The design methodology is similar to the methodology described earlier for the pipelined converters.

The final chip layout was drawn in a semi-automatic manner. The digital section was synthesized from a parametric VHDL code. The layout of the digital circuit was obtained from the available commercial tools in standard cell format. The layouts of the digital block, the comparator bank, and the resistor bank were joined together manually to obtain the final layout, as shown in Figure 5.26. The hardware was Sun Ultra Sparc 5/270 MHz with 128MB RAM. It took two and a half weeks to design the whole circuit.

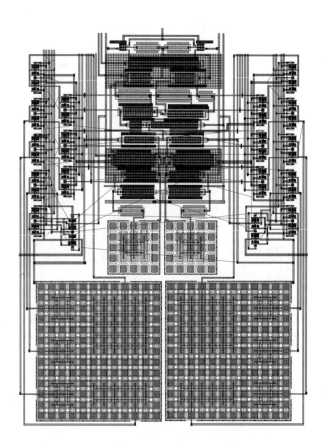

FIGURE 5.25
Layout of the MDAC.

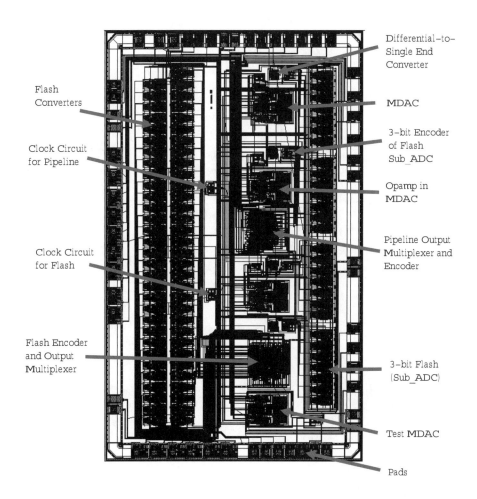

Flash
Converters

Clock Circuit
for Pipeline

Clock Circuit
for Flash

Flash Encoder
and Output
Multiplexer

Differential–to–
Single End
Converter

MDAC

3–bit Encoder
of Flash
Sub_ADC

Opamp in
MDAC

Pipeline Output
Multiplexer and
Encoder

3–bit Flash
(Sub_ADC)

Test MDAC

Pads

FIGURE 5.26
ADC chip layout.

References

[1] Kub, F.J., Moon, K.K., Mack, I.A., and Long, F.M., Programmable analog vector-matrix multipliers, *IEEE Journal of Solid-State Circuits*, 207, 25, 1990.

[2] Gilbert, B., A precise four-quadrant multiplier with subnanosecond response, *IEEE Journal of Solid-State Circuits*, 365, 3,1968.

[3] Shima, T., Kimura, T., Kamatani, Y., Itakura, T., Fujita, Y., and Iida, t., Neuro chips with on-chip back-propagation and/or hebbian learning, *IEEE Journal of Solid-State Circuits*, 1868, 27, 1992.

[4] Steyacrt, M. and Uyrrenhove K., A CMOS 6-bit, 1 GHz ADC for IF Sampling Applications, *IMS2001 International Microwave Symposium*, 2001.

Chapter 6

Conclusion and Future Directions

In this book, we have tried to present several concepts in analog VLSI design automation. Most important of all, we have proposed a design flow for the design automation of analog circuits that takes the problem from the specification stage at the system level and carries it all the way to the layout level, yielding a layout that is ready to be fabricated. This design flow, in turn, specifies the tools to be used in analog design automation. We have clearly listed these specifications and how each tool interacts with the other. We have also developed our own software for each of the tools. Some of the tools we have developed are by no means the best in the literature and can be substituted by better tools as long as the new ones fit into the design flow. However, our work in improving our individual tools is continuing.

Another important point discussed in this book is the working principles of each tool. We have discussed extensively our in-house tools in the corresponding chapters dedicated to them. However, we have by no means neglected reviewing similar tools from the literature and comparing those implementations with ours. Of course, in such a new and exciting field, no literature survey can be complete, and there are bound to be unintentional omissions in the surveys. However, these surveys are intended to be used as a guide for the reader in exploring this field.

In accordance with the above, Chapter 2 presented three system-level synthesis tools. This chapter stressed the high-level models used in the tools more than the optimization approach, since the optimization itself is rather straightforward. The more complicated issue is the development of high-level models that are complex enough to model all important effects, but simple enough to work on. This chapter further defined and discussed the performance estimator. Chapter 3 concentrated mainly on the circuit-level synthesizer. Since this problem is a very difficult optimization problem, the details of the algorithm have been discussed at length, comparing with other available optimizers. Furthermore, some work toward design centering was also presented. Chapter 4 concerns itself mainly with layout synthesis and the layout advisor, where our in-house tool ALG is discussed as well as many examples from the literature.

We have also taken care to present many examples throughout the book. We believe that, in a practical field like engineering, examples are indispensable for following any discussion. The examples presented in this book can be classified into two; examples illustrating the operation of each tool, and complete design examples. The former are scattered throughout the book with several in every chapter, whereas the latter are assembled in a separate chapter. These are treated more as case studies than simple examples where the design procedure, the decisions taken during the design process, and their effects on the final results are discussed in detail.

We believe that we have illustrated the feasibility of the design automation of analog integrated circuits through the discussions and examples presented throughout the book. We do not believe that these tools will replace the insight of the human in the near future – if ever. However, it will make the job of the human designer much easier in many aspects. The designer can use the tools at hand to explore the performances of various architectures at the system level as well as the performances of various standard circuit topologies for a given technology. Having selected a particular system architecture and the topologies of the related blocks, designers can then have the tools synthesize the circuit. They can have the design automation system synthesize the layout or just create directives for critical points in the layout.

Although the whole design automation was designed with a streamlined top-down flow from the specifications to the layout, we are aware that this flow can be desirable only for a novice designer or for a noncritical circuit design. The expert designer, confronted with a difficult design problem, has to consider many intricacies of the design and use his intuition at many points. However, this does not rule out the use of the design automation flow presented here. The tools at every point are well defined and the expert designer may choose to replace one or more of the tools by himself, using only some of them. In addition, the tools have been designed such that interactive operation to some extent is possible in all of them. Thus, the expert designer can use these tools to make design decisions or interact with them during the design. Thus, the design flow is not a static top-down flow, but can be tailored for every application. Digital design experience also shows a similar flow. Although a full top-down flow is also advocated for digital designs, most designers use a hybrid approach where bottom-up concepts are also utilized and the design "meets somewhere in the middle." Just like digital design, analog design rarely flows from top to bottom, but the tools should be able to support this.

Another very important application of analog design automation is technology porting. With technologies being updated almost yearly, analog designers cannot keep up with designing new libraries. Unlike digital design, where libraries can easily be adapted to new technologies, analog libraries have to be redesigned almost from scratch. The only thing that remains intact between libraries is the designer's knowledge of architectures. However, this painful process of porting technologies can become much easier with the use of analog design automation tools. A subject related to this issue is design reuse in the analog domain. Again, the only design that can be reused between technologies and/or institutions in the analog domain is the designer's knowledge, which is best captured with analog design automation tools. We therefore believe that analog VLSI design automation will become a more and more important research subject in the near future.

With this book, we have not covered all areas that pertain to the subject of analog design automation. We have barely discussed macromodeling and symbolic analysis, which are two very important subjects in modeling and optimizing circuits. One reason for this is that these are relatively well established fields with many researchers and a rich literature that the interested reader can consult. Another field that has been briefly mentioned is the field of analog specification languages. This very exciting and promising field should be well studied and considered as an option while developing high-level synthesizers. We have mostly avoided this area, as any discussion and examples would be language-specific, while at this point in time, there is no standard language accepted by most of the community, as in digital design. Furthermore, there is again a fairly large amount of literature discussing these languages and designs employing them. Some subjects have not been even mentioned in the book. One such subject is the testing and design for testability of analog circuits. With automation and increasing complexity in the analog domain, testing is bound to become a bottleneck in the near future and a structured approach is a necessity.

Of the areas discussed in the book, none are static. Quite the contrary, there is a lot of work in accumulating knowledge and finding more efficient and better performing solutions for each problem presented in the previous chapters. The authors are also part of this work and may develop newer and more improved versions of the tools presented here by the time that the book reaches the reader. We thus implore readers to use this book only as a guide for their own further research in this exciting and dynamic field.

Appendix A

CMOS Spice Models

1.5 μm CMOS Technology Hspice Level=49 Model

```
.MODEL CMOSN NMOS (                                          LEVEL    = 49
+VERSION = 3.1          TNOM    = 27              TOX     = 3.19E-8
+XJ      = 3E-7         NCH     = 7.5E16          VTHO    = 0.5449692
+K1      = 0.9382683    K2      = -0.0814687      K3      = 4.9971343
+K3B     = -1.205599    WO      = 1E-7            NLX     = 1E-8
+DVTOW   = 0            DVT1W   = 0               DVT2W   = 0
+DVTO    = 0.5628191    DVT1    = 0.2776207       DVT2    = -0.3521233
+UO      = 670.5086073  UA      = 1.96216E-9      UB      = 1.294827E-1
+UC      = 4.099991E-11 VSAT    = 1.104603E5      AO      = 0.5815491
+AGS     = 0.1394118    BO      = 2.378278E-6     B1      = 5E-6
+KETA    = -8.687558E-3 A1      = 0               A2      = 1
+RDSW    = 3E3          PRWG    = -0.0286019      PRWB    = -0.0370103
+WR      = 1            WINT    = 7.514844E-7     LINT    = 1.890924E-7
+XL      = 0            XW      = 0               DWG     = -1.718509E-
+DWB     = 3.702301E-8  VOFF    = -0.0145079      NFACTOR = 0.9590012
+CIT     = 0            CDSC    = 0               CDSCD   = 9.475364E-6
+CDSCB   = 0            ETAO    = 0.1717744       ETAB    = -0.12651
+DSUB    = 0.899852     PCLM    = 1.3374218       PDIBLC1 = 8.793855E-3
+PDIBLC2 = 2.011998E-3  PDIBLCB = -0.1            DROUT   = 0.0572504
+PSCBE1  = 1.130086E10  PSCBE2  = 2.589851E-9     PVAG    = 0.3117076
+DELTA   = 0.01         RSH     = 52.7            MOBMOD  = 1
+PRT     = 0            UTE     = -1.5            KT1     = -0.11
+KT1L    = 0            KT2     = 0.022           UA1     = 4.31E-9
+UB1     = -7.61E-18    UC1     = -5.6E-11        AT      = 3.3E4
+WL      = 0            WLN     = 1               WW      = 0
+WWN     = 1            WWL     = 0               LL      = 0
+LLN     = 1            LW      = 0               LWN     = 1
+LWL     = 0            CAPMOD  = 2               XPART   = 0.5
+CGDO    = 1.72E-10     CGSO    = 1.72E-10        CGBO    = 1E-9
+CJ      = 2.630847E-4  PB      = 0.99            MJ      = 0.5705142
+CJSW    = 1.442031E-10 PBSW    = 0.99            MJSW    = 0.1
+CJSWG   = 6.4E-11      PBSWG   = 0.99            MJSWG   = 0.1
```

```
+CF        = 0                     )
*
.MODEL CMOSP PMOS (                              LEVEL    = 49
+VERSION = 3.1           TNOM      = 27          TOX      = 3.19E-8
+XJ       = 3E-7         NCH       = 2.4E16      VTHO     = -0.8476404
+K1       = 0.4513608    K2        = 2.379699E-5 K3       = 13.3278347
+K3B      = -2.2238332   WO        = 9.577236E-7 NLX      = 1E-6
+DVTOW    = 0            DVT1W     = 0           DVT2W    = 0
+DVTO     = 1.4951753    DVT1      = 0.451167    DVT2     = -0.0364224
+UO       = 236.8923827  UA        = 3.833306E-9 UB       = 1.487688E-21
+UC       = -1.08562E-10 VSAT      = 1.674377E5  AO       = 0.4435859
+AGS      = 0.1672719    BO        = 5.244897E-6 B1       = 4.525319E-6
+KETA     = 1.765523E-3  A1        = 0           A2       = 0.364
+RDSW     = 3E3          PRWG      = 0.2452639   PRWB     = -0.0635244
+WR       = 1            WINT      = 7.565065E-7 LINT     = 4.875011E-8
+XL       = 0            XW        = 0           DWG      = -2.13917E-8
+DWB      = 3.857544E-8  VOFF      = -0.0877184  NFACTOR  = 0.2508342
+CIT      = 0            CDSC      = 2.924806E-5 CDSCD    = 1.497572E-4
+CDSCB    = 1.091488E-4  ETAO      = 0.26103     ETAB     = -1.34877E-3
+DSUB     = 0.2873       PCLM      = 0.0148934   PDIBLC1  = 6.876272E-4
+PDIBLC2  = 1.006023E-3  PDIBLCB   = -1E-3       DROUT    = 9.918133E-4
+PSCBE1   = 3.339329E9   PSCBE2    = 5E-10       PVAG     = 14.9900194
+DELTA    = 0.01         RSH       = 76          MOBMOD   = 1
+PRT      = 0            UTE       = -1.5        KT1      = -0.11
+KT1L     = 0            KT2       = 0.022       UA1      = 4.31E-9
+UB1      = -7.61E-18    UC1       = -5.6E-11    AT       = 3.3E4
+WL       = 0            WLN       = 1           WW       = 0
+WWN      = 1            WWL       = 0           LL       = 0
+LLN      = 1            LW        = 0           LWN      = 1
+LWL      = 0            CAPMOD    = 2           XPART    = 0.5
+CGDO     = 2.13E-10     CGSO      = 2.13E-10    CGBO     = 1E-9
+CJ       = 3.004422E-4  PB        = 0.7337475   MJ       = 0.4279887
+CJSW     = 1.626626E-10 PBSW      = 0.99        MJSW     = 0.1117313
+CJSWG    = 3.9E-11      PBSWG     = 0.99        MJSWG    = 0.1117313
+CF       = 0                     )             .
*
```

Index

217